Calorimetry: Modern Concepts and Applications

Calorimetry: Modern Concepts and Applications

Edited by **Bruce Rhames**

New York

Published by NY Research Press,
23 West, 55th Street, Suite 816,
New York, NY 10019, USA
www.nyresearchpress.com

Calorimetry: Modern Concepts and Applications
Edited by Bruce Rhames

International Standard Book Number: 978-1-63238-068-5 (Hardback)

Printed in the United States of America.

Contents

Preface

Every book is a source of knowledge and this one is no exception. The idea that led to the conceptualization of this book was the fact that the world is advancing rapidly; which makes it crucial to document the progress in every field. I am aware that a lot of data is already available, yet, there is a lot more to learn. Hence, I accepted the responsibility of editing this book and contributing my knowledge to the community.

As a technique for thermal analysis, calorimetry has a broad spectrum of applications which are not only restricted to studying the thermal characterization (e.g. melting temperature, denaturation temperature and enthalpy change) of small and large drug molecules, but are also extended to characterization of fuel, metals and oils. Differential Scanning Calorimetry is applied to study the thermal behaviors of drug molecules and excipients by calculating the differential heat flow required to maintain the temperature difference between the sample and reference cells. Microcalorimetry is applied to study the thermal transition and folding of biological macromolecules in dilute solutions and could also be applied in formulation and stabilization of therapeutic proteins. This book consists of research from around the world on the usage of calorimetry which is grouped under three sections; namely, Thermal Analysis of Phase Transitions of Polymers and Paraffinic Wax, Indirect Calorimetry to Measure Energy Expenditure, and Applications of Calorimetry into Propellants, Alloys, Mixed Oxides and Lipids.

While editing this book, I had multiple visions for it. Then I finally narrowed down to make every chapter a sole standing text explaining a particular topic, so that they can be used independently. However, the umbrella subject sinews them into a common theme. This makes the book a unique platform of knowledge.

I would like to give the major credit of this book to the experts from every corner of the world, who took the time to share their expertise with us. Also, I owe the completion of this book to the never-ending support of my family, who supported me throughout the project.

Editor

Thermal Analysis of Phase Transitions of Polymers and Paraffinic Wax

Silver Particulate Films
on Compatible Softened Polymer Composites

Pratima Parashar

Additional information is available at the end of the chapter

1. Introduction

Polymer/inorganic nanocomposites have been of great interest in recent years, not only for the novel properties of the nanocomposite materials but also for the continuously growing demand for the miniaturisation of electronics components, optical detectors, chemicals and biochemical sensors and devices.

Polymer matrices have been frequently used as particle stabilizers in chemical synthesis of metal colloids since these prevent agglomeration of the particles. Within the past decade, incorporating silver nanoparticle into a polymer matrix is more interesting because the resulting nanocomposites exhibit applications in catalysis, drug wound dressing and optical information storage.

It is difficult to disperse silver nanoparticle homogeneously into a polymer matrix by ex situ methods because of easy agglomeration of nanoparticles. At present, it is possible to obtain nanoparticles of different shape and size in nanostructured polymeric environment using various polymeric systems and different approaches. Numerous methods used toxic and potentially hazardous reactants. Increasing environmental concerns over synthesis route resulted in an attempt to adopt eco friendly methods.

One of the simplest techniques to form such particulate structures, which are generally known as island or discontinuous metal films, is through vacuum evaporation of metal on to a dielectric substrate by stopping the deposition at a very early stage. The temporal instability exhibited by island films even in vacuum is attributed to mobility of islands followed by coalescence [1]. Further, these films get oxidised when they are exposed to atmosphere. The oxidation of islands causes an irreversible increase in electrical resistance [2]. An interesting sub-surface particulate structure formation was reported when certain inorganic materials are deposited on to softened polymer substrates [3-6] and the morphology and formation of such structures depend on thermodynamic as well as deposition parameters [6, 5]. The use of softened polymer substrats provides the unique possibility of easily controlling the viscosity of the substrate to form a subsurface discontinuous silver particulate films. The morphology of sub-surface particulate structures also depends upon polymer metal interaction [7, 8]. The reported method is evaporation of silver on polymer substrate at high temperature and in vacuum of the order of 10^{-6} Torr. The ability to precisely tailor and optimize the nanocomposite structure creates opportunities for a wide range of applications.

2. Body

Pyridine-containing polymers have attracted interests in recent years because they can be used in various applications as water-soluble polymers and coordination reagents for transition metals, especially 4-vinylpyridine because of its more interesting properties resulting from higher accessibility of the nitrogen atom [9].

Deposition of silver on interacting polymers like Poly (2-vinylpyridine) and Poly (4-vinylpyridine) resulted in the formation of smaller particles (~ a few tens of nm) with smaller inter-particle separations whereas silver deposited on softened inert polymer like polystyrene (PS), irrespective of the deposited thickness is of highly agglomerated structures. Therefore, silver films on inert polymer lack in application due to room temperature resistances equalling that of the substrate. But, silver films on interacting polymers have room temperature resistance in the range of a few tens to a few hundred $M\Omega$/sheet, which is desirable for device applications [8]. Both the interacting polymers are hygroscopic and costly. Therefore, blending an inert and stable polymer like PS with interacting polymers like P2VP and P4VP may provide a polymer matrix suitable for formation of subsurface silver films. Miscibility between the components polymers play a vital role in blending of polymers at the molecular levels. A compatible blend provides a firm basis for further application in devices. Earlier researchers [10-14] have suggested the improvement of miscibility of PS with P4VP by incorporating proton donors like poly (acrylic) acid and poly (p-vinyl phenol) or methacrylic acid into the chains of PS with P4VP in order to utilise its proton acceptor nature. Further, reversible addition-fragmentation chain transfer polymerization was developed by J.J. Yuan and et al [9] for the controlled preparation of PS/P4VP triblock copolymers as PS-b-P4VP-b-PS and P4VP-b-PS-b-P4VP.In order to retain the properties of both the polymers PS and P4VP, blending is carried out through solution casting and it is expected that combination of PS and P4VP should give rise to organised subsurface silver particulate structures with the advantages of both the polymers.

Polymer blending is a common way to develop new polymer materials with desirable combinations of properties. The main advantage of this method is to control the properties by varying the blend compositions [15]. A compatible blend is needed to have desirable combinations of properties of both the polymers. Compatibility of the two homopolymers is needed to an optimum extent for a blend to show superior properties. The compatibility signifies specific interaction such as dipole-dipole, ion-dipole and hydrogen bonding.

Various measurements like heat of mixing, viscometry, glass transition temperature, morphological studies by optical and electron microscopy, infrared spectroscopy and dynamic mechanical analysis, are used to study polymer compatibility. The compatibility of polymer composite is discussed using DSV, DSC, FTIR and SEM. Dilute solution viscometry is a simple and reliable method to investigate interactions of macromolecules in solution. It has been used as a complementary technique to prospect the effect of the position of nitrogen atom in the pyridine ring of P4VP on the interaction developed within PS/P4VP blends. This technique could not be applied to PS/P2VP blends because these blends show phase separation after twenty-four hour of preparation of solution. The criterion of single composition dependent glass transition is used to investigate the miscibility of polymer blends by DSC. Specific interactions most often liberate a heat of mixing and contribute towards the free energy of mixing. Fourier transform infrared spectroscopy is used to investigate specific interactions between the homopolymers in the blend compositions and compared to calorimetric results. SEM results confirm compatibility of blends at higher temperature.

Nanocomposites of metal nanoparticles in a polymer matrix have generated a great deal of interest which depends on the metal-polymer composition and their structure. Polymers are particularly attractive as the dielectric matrix in composites due to their versatile nature and can easily be processed into thin films. These nanocomposites exhibit a unique combination of desirable optical and electrical properties that are otherwise unattainable [16-20].All these properties depend on the size, size distribution and shape of the nanoparticles. The growth and arrangement of metal nanoparticles on various substrates are therefore key issues in all the fields of modern science and technology relating to nanoelectronics, photonics, catalysis and sensors [21].

Since past, polymer membranes have been studied as supporting materials for colloidal metals which are well known catalysis. Strictly speaking, such a membrane contained colloidal metal-rich and metal-poor phases and the localization of colloidal metals is governed by non–linear diffusion equations. Poly (styrene-b-2-vinyl pyridine) diblock copolymer forms micro phase separated film [22] and Ag ion added to such a film is localized in P2VP micro domains not in PS phase. Theoretical study in the self assembly of inorganic/block copolymers hybrids by Ginzburg and co-workers have predicted that affinity, size and amount of inorganic nanoparticles can be exploited to control the phase behaviour of inorganic/BCP hybrids [23].

Metal–polymer nanocomposite containing widely separated nanoparticles exhibit insulating behaviour. As the percentage of metal in composite increases, the nanoparticle separation decreases. At a certain thickness of silver on softened polymer substrate, nanoparticles are quite densely packed but separated by polymer gap such a film offer a host of unique property relevant to practical applications. These applications include high dielectric constant passives, electromagnetic interference shielding, sensors, and detector designed for a variety of specific purposes with high performances, sensitivity and flexibility [24]. Further, the morphology of the cluster films deposited on softened polymer substrates is dependent on the polymer-metal interaction. Gold deposited on polystyrene (PS) and subsequently annealed above its glass transition temperature results in a highly agglomerated film with large separation between the clusters, possibly due to inert nature of PS [1]. Silver deposited on PS also forms highly agglomerated subsurface particulate structure with large separations between the metal particles [25].Also, coalescence rate for gold particles in a poly(2-vinyl pyridine) matrix is much less than the coalescence rate for gold particles in a polystyrene matrix, indicating that homopolymer/metal interactions play an important role in the determination of the coalescence rate [7].Hence, this high coalescence rate in case of PS resulted in a highly agglomerated film even for a thickness of 300 nm. But, silver deposited on an interacting polymer like Poly (2-vinylpyridine) and poly (4-vinylpyridine) resulted in the formation of smaller particles (~ a few tens of nm) with smaller inter-particle separations [8].The differences in dispersion, size distribution and impregnation depth result from the differing natures of the polymer hosts and the processing conditions [26]. Therefore, it is desirable to restrict the nanoparticles to a small size regime along with a narrow size distribution by blending inert PS with interacting P2VP and P4VP.Therefore; it is interesting to blend PS with P2VP and P4VP in order to have desired morphology of silver particulate films on compatible polymer composite.

3. Experimental

Poly (4-vinyl pyridine) and Poly (2-vinylpyridine) used in this study, were procured from Sigma-Aldrich Chemicals Pvt. Ltd and Polystyrene from Alfa-Aesar (A Johnson Mathley company) respectively. The molecular weights of P4VP, P2VP and PS are 60,000, ~37,500 and 100,000, respectively. The structure of P4VP, P2VP and PS are (a), (b) and (c), respectively, as follows:

$[-CH_2-CH-]$ n $[-CH_2-CH-]$ n $[-CH_2-CH-]$ n

(a)   (b)   (c)

Polymer blends were prepared through solution blending by mixing in a common solvent, dimethylformamide (DMF). Blends of PS/P4VP with different compositions {PS (w)/P4VP (w) =0:100; 25:75, 50:50; 75:25; 100:0} were prepared. 1g of the total polymers at different ratios was dissolved in 20 ml of DMF at room temperature. Composite of PS/P2VP with different compositions {PS (w)/P2VP (w) = 0:100; 50:50; 100:0} were prepared. An amount of 0.5 g of the total polymers at different ratios, were dissolved in 5 ml of DMF at room temperature.

The stock solutions of PS, P4VP, and their different blend compositions were prepared in the common solvent DMF. Viscosity measurements were made using Ubbelohde Viscometer at 28°C with an accuracy of ±0.2%.

For DSC study, the solvent is allowed to evaporate in a thermostat for 24 hours. The residuals of component polymers and their blend in powder form were then dried at 80^0 C for several days to ensure complete removal of any traces of residual solvent. The residuals of component polymers and their blends were found to be translucent. DSC measurements were carried out using a Shimadzu DSC-50. Small quantities of the samples, 8-10 mg were scanned at a heating rate of 5- 10 K/min^{-1} in the temperature range 28 to 220^0C under Nitrogen, N_2.

FTIR spectra of the blends were recorded using a Perkin-Elmer spectrometer (model 1000).

Thin films of homopolymers and their composite of approximately $5\mu m$ thickness were solution cast on a glass slide pre-coated with silver contacts with a gap of 1 cm X 1 c m for electrical studies. Silver films of various thicknesses were deposited on these substrates held at 457 K in a vacuum better than 8×10^{-6} Torr. The glass transition temperature of P4VP, P2VP and PS are 410, 357 and 373 K, respectively. Therefore, polymer substrates are softened at 457 K and sufficient fluidity is ensured. A chromel-alumel thermocouple was used to measure the substrate temperature by clamping it to the substrate surface holding the film. Source to substrate distance was maintained at 20 cm. A Telemark quartz crystal monitor (Model 850) was used to measure the deposition rate, as well as the overall film thickness. The deposition rate used was 0.4nm/s for all the films. Resistance measurements were carried out in-situ, using a Keithley electrometer model 617. The films were annealed at the deposition temperature for 1 hour before cooling them to room temperature. Stability of the films against exposure was studied by monitoring the film resistance during exposure to atmosphere by continuously leaking air into the vacuum chamber using a needle valve. The leak rate was such that the pressure increased by an order of magnitude in about a minute.

Optical absorption spectra of the silver particulate films were obtained on a Shimadzu UV-Vis-NIR spectrophotometer model SHIMADZU UV 3101 PC.

Scanning electron microscopy (SEM) measurements were carried out on Scanning electron microscope model JEOL JSM 5800 CV with image processing software.

4. Results and discussion

4.1. Viscosity measurements

The effectiveness of dilute-solution viscometry is based on the assumption that mutual interactions of macromolecules in solution have a great influence on the viscosity in TPS (two polymers in a solvent) [11]. Therefore, compatibility among the polymers depends on the fact that the repulsive interactions among polymer molecules cause their shrinkage, leading to a lowering of solution viscosity, while attractive interaction increases the viscosity.

The relative and reduced viscosities of homopolymer and their blends are found out from viscometric measurements. The intrinsic viscosity values of homopolymers and their blends were determined at 27° C in DMF by extrapolation to zero concentration of the plots of reduced viscosity (η_{sp}/C) versus concentration as shown in figure 1. The plots are not perfect linear, but no crossover is seen. A sharp crossover in the plots of reduced viscosity versus concentration indicates incompatibility of blends [27]. Therefore, some order of compatibility is expected in the blends.

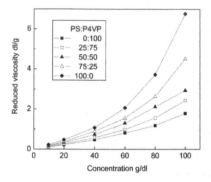

Figure 1. Reduced viscosity versus concentration composition of homopolymers in the blend for the PS/P4VP blends.

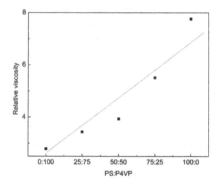

Figure 2. Relative viscosity versus original total concentration of 0.05g/ml.

Figure 2 shows a plot of relative viscosity versus blend composition at the original total concentration of 0.05g/ml. It is not found to be perfect linear for entire range. This indicates that the PS/P4VP blends are not hundred percent incompatible blend system [10, 28, 29].The concentration value is much lower than the critical concentration C″ estimated by C″ =1/ [η] [10].In the absence of specific interactions within the blend, polymer coils are independent if the solution concentration is below the critical concentration [10].The mixed solutions in DMF of PS and P4VP were clear indicating that no strong interactions are taking place between the blend components chains.

As proposed by Krigbaum and Wall [30], the specific viscosity η_{sp} of a solution polymer blends can be expressed as:

$$(\eta_{sp})_m = [\eta_1] C_1 + [\eta_2] C_2 + b_{11}C_1{}^2 + b_{22}C_2{}^2 + 2b_{12}C_1C_2 \tag{1}$$

Where $[\eta_1]$ and $[\eta_2]$ are the intrinsic viscosities of component polymers 1 and 2, C_1 and C_2 are the concentrations of polymers 1 and 2 in solution of polymer blend, b_{11} and b_{22} are specific interaction coefficients of polymers 1 and 2 in single polymer solutions and b_{12} is the interaction coefficient for the polymer blend of component polymers 1 and 2. The coefficient b_{11} is related to the constant k in the Huggins equation, when component polymer 1 is alone in the solution. This also applies to b_{22}.

$$\eta_{sp}/C = [\eta] + k [\eta]^2 C \tag{2}$$

The relationship between b_{11} and k can be written as

$$b_{11} = k_1 [\eta]^2 \tag{3}$$

Where k_1 is the Huggins constant for component polymer 1 in solution. The theoretical interaction coefficient between the two polymers, $b_{12}{}^*$, can be expressed as:

$$b_{12}{}^* = (b_{11} b_{22})^{1/2} \tag{4}$$

According to Krigbaum and Wall [30], information on the intermolecular interaction between polymer 1 and polymer 2 can be obtained by comparison of experimental b_{12} and theoretical $b_{12}{}^*$ values. Hence, the miscibility of binary polymer blends can be characterized by the interaction parameter, Δb:

$$\Delta b = b_{12} - b_{12}{}^* \tag{5}$$

Negative values of Δb are found for solutions of incompatible polymer system, while positive values of Δb refer to attractive interaction in compatible systems. We can reduce the equation (1) to the following form when total concentration of the mixture (C) approaches zero.

$$(\eta_{spm}/C)_{c \to 0} = [\eta_1](C_1/C)_{c \to 0} + [\eta_2](C_2/C)_{c \to 0} \tag{6}$$

Polymer 1-polymer 2 interaction, Δb and theoretically ($\eta_{sp})_m$ can be calculated [15] as follows:

$$(\eta_{sp})_m / C_m = [\eta]_m + b_m C_m \tag{7}$$

Where C_m is the total concentration of polymers, $C_1 + C_2$, and $[\eta]_m$ is the intrinsic viscosity of blend. It can be theoretically defined as;

$$[\eta]_m = [\eta]_1 X_1 + [\eta]_2 X_2 \tag{8}$$

Where X_1 and X_2 are weight fractions of polymer 1 and polymer 2, respectively. Interaction parameter, b_{12}, can be defined by the equation

$$b_m = X_1{}^2 b_{11} + X_1 X_2 b_{12} + 2 X_2{}^2 b_{22} \tag{9}$$

Where b_m defines the global interaction between all polymeric species. b_{12} may be obtained experimentally by Eq. (7).

All the calculated and experimental values are summed up in Table 1. The experimental intrinsic viscosity values are compared with their weighed average values and are found to be lower than the theoretical values. Shih and Beatty [29] have studied immiscible systems by this method and found that the intrinsic viscosity always shows a negative deviation due to the repulsive interaction between the polymers. Hence, these blends were not thermodynamically compatible under equilibrium conditions.

The repulsive deviation causes a reduction in the hydrodynamic volume of the polymer molecules, and hence, the viscosity of the solution is reduced. It is found that Δb values are very much less than unity and negative for all the blends except for the blend 25:75, for which slightly positive value of Δb predicting some order of compatibility. Also, positive deviation in 25:75 can be attributed to increase in the proportion of the polar group, P4VP in the blend [11]. The difference between η_1 and η_2 are found to be large and therefore, a more effective parameter μ can be defined to predict about the compatibility [28].

$$\mu = \Delta b / (\eta_2 - \eta_1)^2 \tag{10}$$

Low values of μ observed in Table 1 may be due to weaker interaction between the polymers. The lower values of interaction parameters indicate that the PS and P4VP are not fully compatible, but physically miscible up to a certain extent.

Blend comp of PS/ P4VP	Intrinsic viscosity		Slope of red viscosity vs. concentration curve	Experimental Value, b_{12}	Theoretical value, b_{12}^*	Δb	μ
	Experimental(dl/g)	Theoretical(dl/g)					
0:100	0.167	0.167	0.018	-	-	-	-
25:75	0.284	0.402	0.024	0.053	0.034		
						0.019	0.021
50:50	0.325	0.638	0.03	0.017	0.034	-	-
						0.017	0.019
75:25	0.64	0.874	0.045	0.018	0.034	-	-
						0.016	0.017
100:0	1.11	1.11	0.067	-	-	-	-

Table 1. Intrinsic viscosity and interaction parametar of PS/P4VP blends.

4.2. Differential scanning calorimetery

DSC endothermograms for the homopolymers and their blends are shown in figure 3

All the blends exhibit a single T_g, intermediate between those of the parent polymers, PS and P4VP indicating the miscibility of these blends. The theoretical values of these can be predicted using Fox equation [31] and Gordon-Taylor equation [32].

$$1/T_g = X_1/T_{g1} + X_2/T_{g2} \tag{11}$$

$$T_g = (X_1 T_{g1} + k X_2 T_{g2})/ (X_1 + k X_2) \tag{12}$$

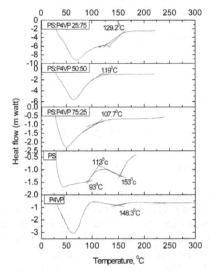

Figure 3. DSC thermograms of PS/P4VP blends

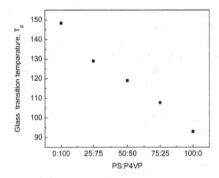

Figure 4. Glass transition temperature versus composition of PS/P4VP.

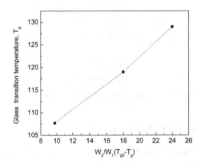

Figure 5. Verification of Gordon- Taylor equation for PS/P4VP blends.

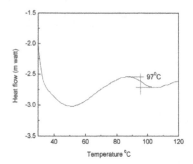

Figure 6. DSC thermogram of PS/P2VP, 50:50

Where X_1, X_2, T_{g1} and T_{g2} are the weight fractions and glass transition temperatures corresponding to polymer 1 and polymer 2, respectively. k is a constant which gives a semi-quantitative measure of degree of the interaction between the two polymers. All the experimental and calculated values of Tg, are shown in Table 2. Positive deviation observed from Fox equation is attributed to intermolecular interaction between the polymers. Figure 4 shows the plot of T_g with blend composition. It is well established that when interactions between blend components are strong, such as those affected by Hydrogen bonding, the experimentally determined T_g of the blends are higher than those calculated from the additivity rule as a result of the reduction of polymer chains mobility in the blend [10]. In order to estimate the strength of the intermolecular interactions within the PS/P4VP blends, we used the Gorden-Taylor equation to verify through the linear fit in figure 5. Slope (k) of the straight line obtained is found to be 0.85, indicating interaction between the polymers [32]. The intercept is about 100.47°C which corresponds to T_g of pure PS.

Blend comp of PS/P4VP	Experimental T_g value (°C)	Theoretical T_g value(°C)
	DSC	Fox equation
0:100	148.3	-
25:75	129	125
50:50	119	116
75:25	107.7	107
100:0	93	-

Table 2. Experimental and theoretical glass transition temperatures of PS/P4VP blend.

Figure 6 shows DSC thermogram of the blend PS/P2VP, 50:50.This indicates a single T_g, about 370K intermediate between those of the parent polymers, PS and P2VP indicating the compatibility of the blend on melt mixing at higher temperature.

4.3. Fourier transform infrared spectroscopy

The proton donor PS copolymer, PSMAA (PS-methacrylic acid) with P4VP in solvent chloroform observed the specific interactions with the formation of hydrogen bonds, at a frequency, v = 1607 cm⁻¹ corresponding to the –COOH…N≤ hydrogen bond [11]. Therefore, the incorporation of MAA into PS results in an augmentation in miscibility with P4VP, in comparison with the PS/P4VP system, which is highly incompatible [11]. The present solution cast PS/P4VP blends, needs to be applied in thin film form at higher temperature around 200⁰ C. Therefore, we need to ascertain about their compatibility at higher temperature. Figure 7 is the FTIR of the sample of the polymer blend PS/P4VP, 50:50 before and after DSC being carried out. This shows the absorption bands at 1598 and 1414 cm⁻¹ corresponding to the pyridine ring of P4VP and at 822 cm⁻¹ to the single substituted pyridine appeared in the spectra of PS/P4VP blend. Similarly, for PS, the absorption peaks at 1493cm⁻¹ and 1448 cm⁻¹, which were characteristic of the phenyl ring and the peak at 697cm⁻¹, corresponding to the signals of the single substituted phenyl ring, appeared for the blend as well. Similar trends were observed by the triblockpolymers PS-P4VP-PS and P4VP-PS-P4VP, which were synthesized by chain transfer agent [33]. In case of PS-block-P4VP, the stretching bands overlap [14].The silver particulate film deposited on PS/P4VP (50:50) resulted in desired structure underlying the property of PS and P4VP [34].Figure 7 shows no shift in the frequency leads to the absence of hydrogen bond. Thus, the possibility of protonation of nitrogen of P4VP [14] is ruled out but some intermolecular interaction at higher temperature leads to single Tg composition.

Figure 8 shows FTIR of PS/P2VP, 50:50 after DSC which clearly indicates the absorption bands at 1594 and 1414 cm⁻¹ corresponding to the pyridine ring of PS and at 822 cm⁻¹ to the single substituted pyridine ring appeared in the spectra of PS/P2VP blend. Similarly, the absorption peaks at 1495cm⁻¹ and 1448 cm⁻¹, which were characteristic of the phenyl ring and the peak at 697cm⁻¹, corresponding to the signals of the single substituted phenyl ring for P2VP spectra, also appeared in the spectra of PS/P2VP, 50:50 blend. No shift in the frequency ruled out the possibility of any hydrogen bond in PS/P2VP. But single T_g of the blend suggests some order of the compatibility at higher temperature.

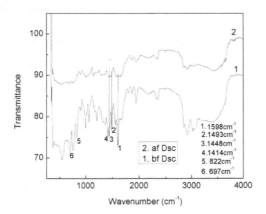

Figure 7. FTIR for PS/P4VP (50:50) before and after DSC.

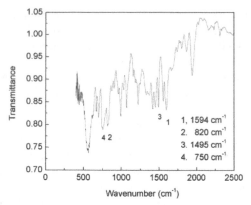

Figure 8. FTIR of PS/P2VP, 50:50 after DSC

4.4. Scanning electron microscopy

Fig. 9a and 9b show the SEM of PS: P4VP (50:50) samples after and before the DSC have been carried out. It is clear that after DSC the blend mixed better than the blend as obtained at room temperature by solution cast which eventually show phase separation after few days of preparation. Fig.9b clearly shows the dispersed phase of polymers whereas Fig. 9a indicates better compatibly of polymers. This can be attributed to mixing of homoplymers around 200°C during the process of DSC. This is an indication of suitable compatibility of these blends at higher temperature.

Figs. 10a and 10b are the Scanning Electron Micrographs of PS/P2VP, 50:50 samples before and after the DSC have been carried out. It is clear that melt mixing at higher temperature gives more compatible blend than room temperature solution mixed blend. Thus, an order of compatibility is achieved in PS/P2VP, 50:50 blend as reported for PS/P4VP blends [35]. Therefore, we can expect formation of discontinuous silver subsurface film on the blend.

4.5. Electrical behaviour of discontinuous silver films on PS/P2VP and PS/P4VP

Figure 11&12 shows the variation of the logarithm of resistance against inverse of temperature for silver films of different thicknesses deposited on polymers and their composite at a temperature of 457 K, during cooling to room temperature. It is interesting to note that while some of the films show only negative temperature coefficient of resistance (TCR) some show almost zero TCR. Some of the films show negative TCR at higher temperatures and almost zero TCR at lower temperatures. The 50 nm thick silver films on pure PS and 75:25 blend of PS/P4VP show negative TCR. Silver on PS showed similar behaviour in our earlier studies resulting in room temperature resistance same as that of the substrate with the formation of large silver particles separated by large distances [25]. Blending the inert polymer PS with an interacting polymer like P4VP to the extent of 25% does not seem to alter the morphology of the particulate film as indicated by the electrical

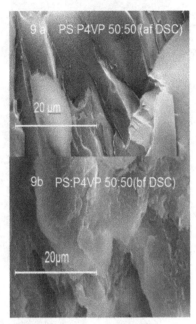

Figure 9. SEM photographs of PS/P4VP (50:50) a. after DSC and b. before DSC.

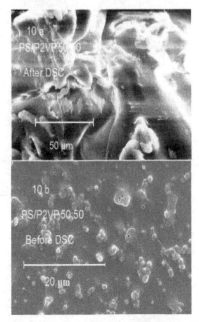

Figure 10. SEM photographs of PS/P2VP, (50:50) (a) after and (b) before DSC.

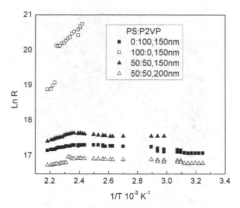

Figure 11. Variation of ln R with 1/T for silver films deposited on the composite PS/P2VP held at 457 K at a rate of 0.4 nm/s.

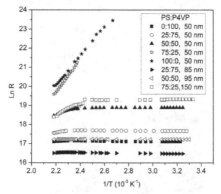

Figure 12. Variation of ln R with 1/T for silver films deposited on the composite PS/P4VP held at 457 K at a rate of 0.4 nm/s.

behaviour. When the P4VP content is increased to 50%, a negative TCR at high temperature followed by almost zero TCR at lower temperatures exhibited by the 50 nm thick film is similar to the behaviour observed earlier for the case of pure P2VP and P4VP [36, 37] indicating that the film consists of small particles separated by small distances. With further increase in P4VP content, the negative TCR part diminishes, giving rise to a near zero TCR .Similarly, the films on composites PS/P2VP (50:50) initially show negative TCR but zero TCR at lower temperature. Blending of PS with P2VP and P4VP seems to result in a positive effect on the composites. Therefore, negative TCR is totally vanishing and give rise to near zero TCR at room temperature for films deposited on the composites. The over all resistance of film deposited on composite decreases with increase in thickness of silver films deposited on the composite. Similar trend were reported that the electrical conductivity of composites is increased with high silver loading (30-80%) [38]. It is also interesting to note that even at 50% P4VP, with an increase of silver deposited, films show electrical characteristics as that of the films on pure P4VP [37]. Further, when 150 nm thick silver is deposited on a polymer blend with only 25% of P4VP, the films show desirable electrical characteristics in contrast with the very high room temperature resistance observed for films on pure PS even at 300 nm of silver [25].

Figure 13&14 show the variation of logarithm of resistance (ln R) with logarithm of pressure in torr [ln (pressure)]. It is seen that the resistances show large increase beyond a pressure of about 0.5 torr, for all the films. The variation in resistance is very small till that pressure. Similar characteristics were shown by silver films on softened P4VP [39] and P2VP [36]. It was shown through X-ray photoelectron spectroscopy (XPS) studies at various electrons take off angles that silver clusters are formed at a depth of a couple of nm from the polymer surface [36,37]. It is known that the formation of subsurface particulate structure formation is subject to certain thermodynamic [6] and deposition conditions [5]. While the thermodynamic conditions are met for the deposition of metals on most of the polymer substrates, deposition conditions used in the present study are similar to those used in our earlier studies. Therefore, it is reasonable to assume that the particles are formed just a couple of nm below the polymer surface. The behaviour of the particulate films upon exposure to atmosphere is attributed to oxidation of islands due to the inadequate polymer cover.

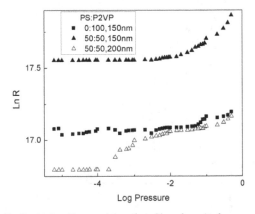

Figure 13. Variation of ln R with log (Pressure) for silver films deposited on composite PS/P2VP held at 457 K at a rate of 0.4 nm/s.

Figure 15 shows variation of room temperature resistance of silver particulate film of 150 nm thickness on PS/P2VP blends .It clearly indicates the decrease in resistance with increase in the amount of P2VP in the blend. Figure 16 shows variation of room temperature resistance of silver particulate film of various thicknesses on the blend PS/P2VP, 50:50.With the increase in the thickness of the silver particulate film the room temperature resistance of the film decreases [34]. Blending of P2VP with PS is expected to provide a polymer matrix where the size of silver clusters and inter-cluster separation can be modified because dispersion, size distribution and impregnation depth results from the natures of polymeric hosts [40].

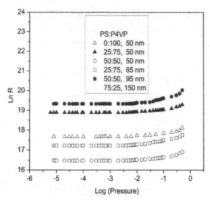

Figure 14. Variation of ln R with log (Pressure) for silver films deposited on composite PS/P4VP held at 457 K at a rate of 0.4 nm/s.

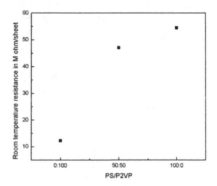

Figure 15. Variation of room temperature resistance of silver particulate film of 150 nm thickness verus composition of PS/P2VP.

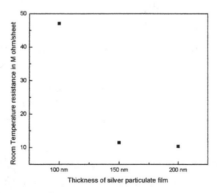

Figure 16. Variation of room temperature resistance of silver particulate film verus their thicknesses for PS/P2VP,50:50.

Figure 17 a,b,c shows the SEM of silver particulate films of thickness 200 nm on PS, P2VP and P2VP, 50:50.It is evident from the figure that decrease in the particle size in the films of P2VP and PS/P2VP, 50/50 resulted in the close proximity of particles with reduction in the inter-particle separation. Also, the decrease in the size of silver cluster in the blend PS/P2VP, 50:50 improves the tunnelling effect as expected [34].Thus, the room temperature resistance of silver particulate film on the blend is now in the desirable range for applications.

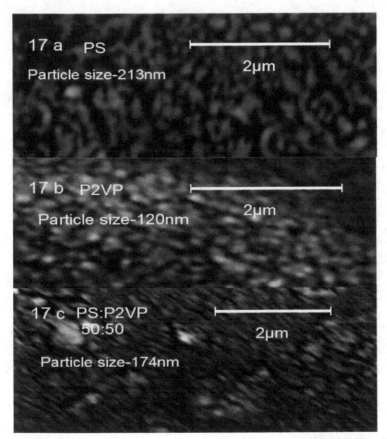

Figure 17. a,b,c: SEM of silver particulate films of thickness 200 nm on PS, P2VP and PS/P2VP, 50:50.

Figure 18 gives the variation of as deposited and room temperature resistances as a function of PS/P4VP composition for silver films of 50 nm thickness. It is seen that as the P4VP concentration increases there is a regular decrease of resistance at a fixed silver thickness. The plot of logarithm of these resistances with blend concentration gives linear fit as shown in figure19.Through this fit, one can estimate the resistance of the film at a particular blend and for the given conditions and thickness.

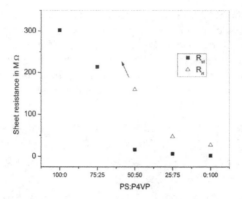

Figure 18. Variation of as-deposited and room temperature resistance with PS/P4VP blend composition for 50 nm silver films deposited at 457 K.

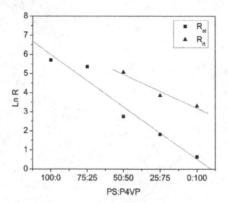

Figure 19. Variation of logarithm of resistances at 457 K and at room temperature with PS/P4VP blend composition for 50 nm silver films.

Table 3&4 gives the resistance data for the silver films of different thicknesses deposited on the PS/P2VP and P4VP, respectively.

Polymer PS:P2VP	Silver Film Thickness	Resistances (MΩ/sheet)				
		R_{ts}	$R_{1\ hra}$	R_{rt}	$R_{0.5T}$	R_{atm}
0:100	100 nm	12.28	32.2	32.68	34.22	63.4
0:100	150 nm	9.73	28.08	26.13	30.53	45.64
50:50	100 nm	47.07	111.57	114.76	489	>1000
50:50	150 nm	11.5	34.98	42.04	57.4	76.56
50:50	200 nm	10.35	29.9	30.1	38.41	68.25
100:0	150 nm	54.46	159.3	-	-	-

Table 3. Resistances for silver films deposited on PS/P2VP blends held at 457 K with a rate of 0.4 nm/s.

Polymer PS:P4VP	Silver film thickness	Resistances (MΩ/□)				
		R_{st}	$R_{1\ hra}$	R_{rt}	$R_{0.5t}$	R_{atm}
0:100	50 nm	1.9	26.8	27.1	32.8	72.3
25:75	50 nm	6.2	42.2	47.1	73.7	1042
25:75	85 nm	3.2	14.7	14.2	21.9	215
50:50	50 nm	15.9	119.5	159.4	241.3	2465
50:50	95 nm	2.9	29.9	30.1	52.2	418
75:25	50 nm	214	325	-	-	-
75:25	150 nm	14.8	98.2	248.9	501.3	5172
100:0	50 nm	302	491	-	-	-

Table 4. Resistances for silver films deposited on PS/P4VP blends held at 457 K with a rate of 0.4 nm/s.

4.6. Morphology of silver particulate films on PS/P2VP and PS/P4VP blends

4.6.1. Optical studies

Fig. 20a shows the optical absorption spectra recorded for 100 nm silver films deposited on polymeric blends of PS/P2VP held at 457 K at the deposition rate of 0.4 nm/s. It is well known that small silver particles embedded in polymer matrix exhibit plasmon resonance absorption and as a result absorption maxima occur in the visible-near infrared region and their spectral position depends on the particle size, shape, filling factor etc. in the polymer matrix. The surface plasmon resonance absorption for silver clusters in the polymer matrix generally occurs at a wavelength of ~ 430 nm [8]. It is well known that shift in the plasmon resonance peak towards higher wavelength occurs due to close proximity of the silver clusters [41-44].These nanoparticles exhibit unique optical properties originating from the characteristic surface plasmon by the collective motion of conduction electrons [43,44].Thus, the formation of silver nanoparticles can also be confirmed by UV/VIS absorption spectrum of composite films [44].Spectral position, half width and intensity of the plasmon resonance strongly depend on the particle size, shape and the dielectric properties of the particle material and the surrounding medium [45].Thus, the type of metal and the surrounding dielectric medium play a significant role in the excitation of particle plasmon resonance (PPR). The sensitivity of PPR frequency to small variations of these parameters can be exploited in various applications [46]. The differing natures of the polymeric hosts yield change in dispersion, size distribution and impregnation depth of silver clusters [26]. Therefore, silver particles embedded in PS/P2VP blends, a shift of the resonance position to higher wavelength (red shift) were found, which were correlated with changes of particle sizes and inter-separation in silver clusters. It is clearly seen (Fig.20a) that the plasmon resonance peak shifts towards the longer wavelength side for the PS/P2VP, 50:50 (435) as compared to pure polystyrene (429 nm). Also, there is increase in intensity of absorbing peaks which signify the decrease of particles size with the incorporation of P2VP into PS [44].P2VP exhibits two peaks (441,606 nm). It is interesting to note that PS/P2VP, 50:50 also shows an additional absorption band at higher wavelength (616 nm). The possible explanation is that silver nanoparticles are in a highly aggregated state leading to coupling of the plasmon vibrations between neighbouring particles [47].Similar results were found for silver particulate films of 150 and 200 nm films on PS/P2VP 50:50 blends. The shift in plasma resonance towards higher wavelength indicates close proximity and increase in particle size of silver nanoparticles with increasing thickness of silver particulate films [41].

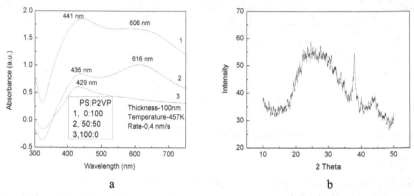

a   b

Figure 20. a: Optical absorption spectra for 100 nm silver particulate films deposited on PS/P2VP blends, **b**: XRD curve for silver particulate films of thickness 200 nm on PS/ P2VP, 50:50.

Fig.20 b shows the XRD pattern with the diffraction peak around 38⁰ for the silver particulate films of thickness of 200 nm on PS/P2VP, 50:50. The broadening of the Bragg peaks indicates the formation nanoparticles. The particle sizes calculated from the Fig.20b is 40.6 nm for the silver particulate film of 200 nm on PS: P2VP, 50:50. The particle sizes estimated from XRD suggest that there is a small reduction in particle size due to blending of PS and P2VP. XRD of PS/P2VP, 50:50 for 200 nm has been carried out to have preliminary idea about average size of the silver clusters.

Fig. 21(a) shows the optical absorption spectra recorded for 50 nm silver films deposited on polymeric blends of PS/P4VP held at 457 K at the deposition rate of 0.4 nm/s. For silver particles embedded in PS/P4VP blends, a shift of the resonance position to higher wavelength (red shift) were found, which were correlated with changes of particle sizes and inter-separation in silver clusters. It is clearly seen (Fig.21a) that the plasmon resonance peak shifts towards the longer wavelength side in comparison to pure PS (485.5 nm).It is 598, 651.6 and 754.5 nm for PS: P4VP, 75:25, 50:50 and 25:75, respectively for 50 nm silver particulate films on them. Also, there is increase in intensity of absorbing peaks which can be attributed to the decrease of particles size with the incorporation of P4VP into PS [48].

Fig. 21 (b) shows the optical spectra recorded for the films of various thicknesses on PS/P4VP blends. The blend 50:50 shows the most promising result among all the silver particulate films on the blends. The intensity and shift of absorption peak is optimum (793 nm) for 95 nm film on PS/P4VP, 50:50.The silver particulate film of thickness 150 nm on PS/P4VP, 75:25 also shows a red shift (705 nm).This shift in the plasmon resonance peak towards higher wavelength can be attributed to close proximity of the silver clusters [42-45], perhaps this is the reason for the better electrical behaviour of the film of thickness 150 nm on PS/P4VP, 75:25 [34].

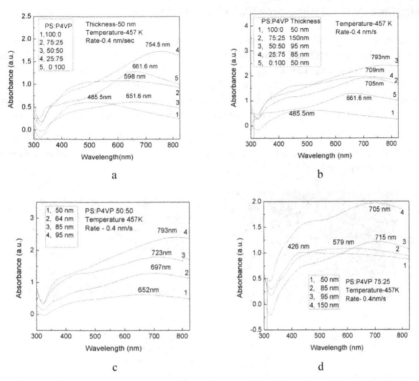

Figure 21. a Optical absorption spectra for 50 nm silver particulate films deposited on PS/P4VP blends, b Optical absorption spectra for the films of various thicknesses deposited on the PS /P4VP blends, c Optical absorption spectra for the films deposited on the blend PS/P4VP 50:50, d Optical absorption spectra for the film deposited on the blend PS/P4VP 75:25.

Fig. 21 (c) shows the optical spectra for films of varying thickness on PS/P4VP, 50:50.The shift in plasma resonance towards higher wavelength indicates close proximity of silver nanoparticles with increasing thickness of silver particulate films [15].

Fig. 21(d) shows the optical absorption spectra recorded for silver films on PS/P4VP, 75:25. It is clear that shift in surface plasma resonance is due to increase in particle size with the amount of silver deposited [8].The results are in agreement with the electrical properties of this blend which show increase in electrical conductivity on increasing silver loading [34].

As the fraction of the metal in a nanocomposite increases the nanoparticle separation decreases resulted in better electrical properties of nanocmposites [49]. Therefore, electrical behaviour of silver particulate films on PS/P4VP (50:50, 50 nm and 95 nm; 75:25, 50 nm and 150 nm) observed decrease in electrical resistance on increasing the thickness of film

[40].Thus, electrical studies of these blends suggest possibility of modification in morphology of silver particulate films on PS/P4VP as compared to films on PS.

Previous studies [8] of silver particulate films on PS for the 100 nm thickness exhibited minimum shift due to the presence of comparatively larger clusters with larger inter-cluster separations than on P2VP [8] for the same thickness. Also, silver particulate film on PS for the 50 nm thickness exhibited minimum shift due to the presence of comparatively larger clusters with larger inter-cluster separations than on P4VP [37] for the same thickness. Blending of P2VP with PS and P4VP with PS is expected to provide a polymer matrix where the size of silver clusters and inter-cluster separation can be modified because dispersion, size distribution and impregnation depth results from the natures of polymeric hosts [40].

Therefore, shift in the wavelength observed in optical spectra in the PS/P2VP, 50:50, and PS/P4VP, 75:25; 50:50; 25:75 can be attributed to modification in size distribution and better inter-cluster separations.

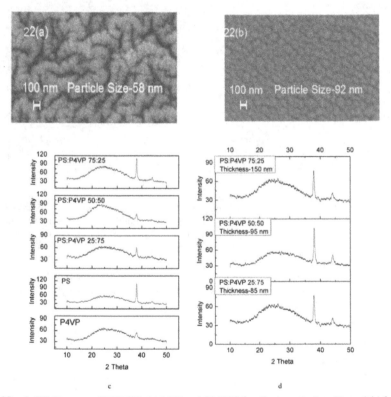

Figure 22. a,b: SEM images of PS/P4VP, (a) 0:100 and (b) 100:0 for silver particulate films of thickness 50 nm. Acceleration voltage- 20 kV, Magnification 50 KX, **c**: XRD curves for silver particulate films of thickness 50 nm on P4VP, PS, PS/ P4VP, 25:75, 50:50, 75:25, respectively. **d**: XRD curves for silver particulate films on PS/P4VP, 75:25, 50:50 and 25:75 for 150, 95 and 85 nm thicknesses, respectively

Fig.22c shows the XRD pattern with the diffraction peak around 38^0 for the silver particulate films of thickness of 50 nm on P4VP, PS and PS/P4VP, 25:75, 50:50, and 75:25. The broadening of the Bragg peaks indicates the formation nanoparticles. The particle sizes calculated from the Fig.22c are 53, 51, 49, 46 and 30 nm for the blends (PS: P4VP) 100:0, 75:25, 50:50, 25:75 and 0:100, in that order. The particle sizes estimated from XRD suggest that there is a very small reduction in particle size due to blending of PS and P4VP.Figure 22d shows the XRD patterns for the silver particulate films on PS/P4VP, 75:25, 50:50 and 25:75 for 150, 95 and 85 nm thicknesses, respectively. The reflections at 38^0 and 44^0 correspond to metallic silver. The particle sizes calculated from the Fig.22d are 52.3, 51.6 and 53 nm for the blends (PS: P4VP) 75:25, 50:50, and 25:75, respectively for the diffraction angle 38^0.

4.6.2. Micro structural studies

Electrical properties of polymer/metal composite films are strongly linked to particles' nanostructure [40]. As the fraction of the metal in a nanocomposite increases the nanoparticle separation decreases resulted in better electrical properties of nanocmposites [49-50,52,53]. SEM of the silver particulate films (Fig.22a & 22b) on homopolymers (PS, P4VP) clearly shows the characteristic nature of these polymers. The size and inter-separation of silver clusters is less (average particle size-58 nm) in P4VP whereas size as well as inter-separation is wide in PS (average particle size-92 nm). As a result, PS do not show the desired electrical conductivity [51].Blending of P2VP and P4VP into PS modifies size, size distribution and inter-separation of silver particles deposited on their blends PS/P2VP, 50:50 and PS/P4VP, 50:50 and 75:25. Therefore, electrical behaviour of silver particulate films on PS/P2VP, 50:50 for 100, 150 and 200 nm observed decrease in electrical resistance on increasing the thickness of films [52]. Thus, electrical studies of these blends suggest possibility of modification in morphology of silver particulate films on PS/P2VP and PS/P4VP as compared to films on PS.

SEM of the silver particulate films of 200 nm on the homopolymers (PS, P2VP) and their blend PS/P2VP, 50:50 were shown in the Fig.23a. The acceleration voltage is 30 kV and magnification is 60 to 100 KX for all the SEM pictures. The particle sizes measured from respective SEM pictures are plotted as histogram. The corresponding histograms (Fig.23a) of silver particles of the films are shown side by side of the SEM pictures. The data fit into a log normal distribution for all the cases. Hence, the average size, $\bar{a}$ and geometric standard deviation, σ a are determined from the log normal distribution of the curves. The average size, $\bar{a}$ and geometric standard deviation, σ a are 205 nm and 4; 109 nm and 9, 129 nm and 3, respectively, for silver films on PS, P2VP and their blend (50:50). The size distribution and width of histograms as shown in the figure indicate that the particle size varies from 100 to 400 nm, 60 to 180 nm and 60 to 200nm for silver films on PS, P2VP and their blend (50:50).It is clear from the figure that the size and inter-separation of silver clusters is wide in PS whereas size as well as inter-separation is less in P2VP [8]. As a result, PS do not show the desired electrical conductivity [36].Blending of P2VP with PS modifies size, size distribution and inter-separation of silver particles on their blend PS/P2VP, 50:50 which results in improvement of tunnelling effect in the blend and the blend shows desired electrical behaviour [52]. This fact may be regarded as a consequence of the size as well as inter-separation evolution of nanoparticles during the ongoing deposition process.

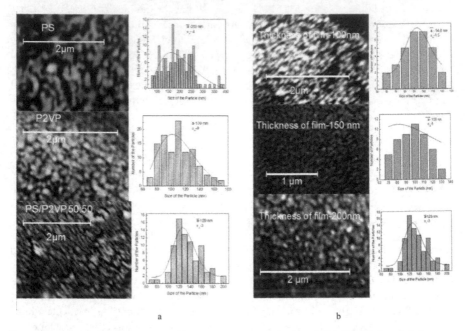

a   b

Figure 23. a: SEM of silver particulate films of 200 nm on PS, P2VP and PS/P2VP, 50:50 and their corresponding histograms. **b**: SEM of silver particulate films of 100, 150 and 200 nm on PS/P2VP, 50:50 and their corresponding histograms.

Fig.23b shows the particle size distribution for 100, 150 and 200 nm thickness films deposited on PS/P2VP, 50:50 and their corresponding histograms. The average size, $\bar{a}$ and geometric standard deviation σ_a are 94.8 nm, 1.5 and 100 nm, 5 and 129 nm, 3 for the 100, 150 and 200 nm film, respectively. It is evident from the figure that size distribution varied from 55-125, 65-135 to 70-200 nm for the 100, 150 and 200 nm films, respectively. Such dispersion of silver nanoparticle within the PS/P2VP, 50:50 leads to better electrical behaviour [52].This electric behaviour is not observed even for 300 nm silver particulate films on PS [20].Hence, silver particulate films on PS/P2VP, 50:50 at low volume fraction of silver consist of isolated and widely dispersed nanoparticles. But systematic and controlled increase of silver volume in the blend matrixes has shown increase in the size of silver clusters [52-53].

Figs.24a to 27a show the SEM pictures of various thicknesses of silver films deposited on PS/P4VP blends. The acceleration voltage is 20 kV and magnification is 50 to 100 KX for all the SEM pictures. The particle sizes measured from respective SEM pictures are plotted as histogram in figs. 24b to 27 b. The corresponding histograms (Figs.24b to 27 b) of silver particles of the films are shown side by side of the SEM pictures. The data fit into a log normal distribution for all the cases. Hence, the average size, ā and geometric standard deviation, σ ₐ are determined from the log normal distribution of the curves. The positive effect of blending P4VP with PS is clearly visible in these pictures. Figs. 24a, 25a show the particle size distribution for 50 and 95 nm thick silver films deposited on PS/P4VP, 50:50. The average size, ā and geometric standard deviation, σ ₐ are 79.4 nm and ±1.2, respectively for the 50 nm film whereas the corresponding values for the 95 nm film are 95.4 nm and ±1.4. A closer look at the morphology of silver nanoparticles deposited on PS/P4VP, 50:50 in Figs.24a & 25a, clearly shows that particle size increases with the amount of silver deposited. The shape of the nanoparticles changes from near spherical particles to irregular ellipsoidal particles. The size distribution and width of histograms as shown in the figure indicate that the average size of the particle increases from 79.4 to 95.4 nm and the size distribution expands from 50-110 nm to 60-160 nm which results in improvement of tunnelling effect in PS/P4VP, 50:50 [40].And silver particulate film of thickness 95 nm show better electrical behaviour than silver particulate film of 50 nm on PS/P4VP, 50:50 [34]. It is evident that increase in size distribution decreases the inter-separation of silver clusters in this blend. This fact may be regarded as a consequence of the size as well as inter-separation evolution of nanoparticles during the ongoing deposition process.

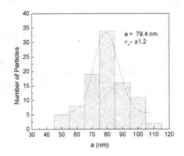

Figure 24. a SEM micrograph, Acceleration voltage-20 kV, Magnification 50 KX and b Corresponding histogram of 50 nm thick silver film on PS/P4VP, 50:50.

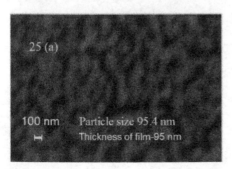
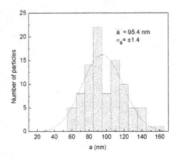

Figure 25. a SEM micrograph, Acceleration voltage-20 kV, Magnification 100 KX and b Corresponding histogram of 95 nm thick silver film on PS/P4VP, 50:50.

Figs.26a & 27a show the particle size distribution for 50 nm and 150 nm thickness films deposited on PS/P4VP, 75:25. The average size, ā and geometric standard deviation, σ a are 88.6 nm and ±8.5 for the 50 nm film and 96.7 nm and ±4 for 150 nm film, respectively. It is evident from the figure that distribution of size increased from 50-120 nm to 60-140 nm .Such dispersion of silver nanoparticle within the PS/P4VP, 75:25 leads to better electrical behaviour [40].Such electric behaviour is not observed even for 300 nm silver particulate film on PS [36].Hence, silver particulate film of 50 nm thickness on PS/P4VP, 75/25 consist of isolated and widely dispersed nanoparticles. But systematic and controlled increase of silver in the PS/P4VP, 75:25 matrixes produced interesting result [40]. The silver particulate film of thickness 150 nm on PS/P4VP, 75:25 shows the room temperature resistance in few mega ohms a desirable range for applications.

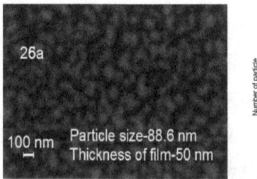
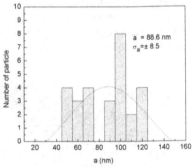

Figure 26. a SEM micrograph, Acceleration voltage-20 kV, Magnification 100 KX and **b** Corresponding histogram of 50 nm thick silver film on PS/P4VP, 75:25.

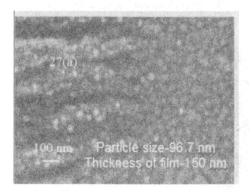
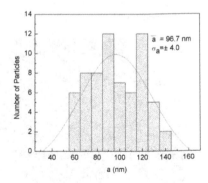

Figure 27. a SEM micrograph, Acceleration voltage-20 kV, Magnification 100 KX and **b** Corresponding histogram of 150 nm thick silver film on PS/P4VP, 75:25.

Table 5&6 has been compiled to show all the particles size of silver clusters embedded in PS/P2VP and PS/P4VP blends from XRD and SEM. The SEM has provided the morphology of silver clusters and their distribution. The XRD diffraction pattern represents the average throughout the film due to increased penetration and large beam size. The observed values of particle size from SEM are in the same range as the calculated values of the particle size from XRD. The difference in the values may be due to averaging over longer depths because of penetration of X-rays. Though, the trend of particle size measured from XRD and SEM are similar.

PS/P2VP	Thickness (nm)	Particle size	(nm)	Standard deviation σ_a
		XRD	SEM, $\bar{a}$	
0:100	200		109	9
50:50	100	-	94.8	1.5
	150	-	100	5
	200	40.6	129	3
100:0	200		205	4

Table 5. The average particle size for silver deposited on PS/P2VP blends. The deposition rate is 0.4nm/s and temperature is 457K.

PS/P4VP	Thickness (nm)	Particle size	(nm)	Standard deviation σ_a
		XRD	SEM, $\bar{a}$	
0:100	50	30	58	-
25:75	50	46.6	-	-
	85	53	-	-
50:50	50	49.1	79.4	1.2
	95	51.6	95.4	1.4
75:25	50	51.8	88.6	8.5
	150	52.3	96.7	4.0
100:0	50	53.3	92	-

Table 6. The average particle size for silver deposited on PS/P4VP blends. The deposition rate is 0.4nm/s and temperature is 457K.

5. Further research

The silver films deposited on polymer composites show an increase in resistance, when they are exposed to atmosphere. It may be possible to stabilise the resistance against exposure to atmosphere through the deposition of good quality inorganic passivators like alumina, zirconia etc, to make the films suitable for device applications. Also, silver films are more susceptible to oxygen in atmosphere than gold. Hence, deposition of gold nanopaticles on PS/P2VP and PS/P4VP composites may give more stabilised films which may be suitable films for sensitive applications.

6. Conclusion

The following conclusions may be drawn from the study on **Silver Particulate Films on Compatible Softened Polymer Composites**

Viscometry studies indicate a very small interaction parameter resulting in physically miscible blends of PS/ P4VP.DSC studies indicate a single T_g in all the cases indicating the formation of compatible blends. This may be due to some intermolecular interaction at higher temperature. Hence, the blends found some order of compatibility at higher temperatures. FTIR and SEM support the results of miscibility as well as DSC. The fairly compatible blend of PS/P2VP and PS/P4VP can be obtained on melt mixing at higher temperature. Deposition of silver on polymer blends coated substrate held at 457 K provides an approach to produce stable island films with reasonable control over their electrical resistance. Higher thickness films show almost zero TCR near room temperature, a desirable property for most of the devices. Low thickness films show a negative TCR, characteristic of island film. Silver particulate films deposited on composite blends show better electrical properties compared to pure PS.The deposition of silver particulate films by evaporation on PS/P2VP and PS/P4VP blends yields positive effect of blending PS with P2VP and P4VP.The size distribution and dispersion of silver nanoparticles is found to be dependent on the nature of the polymer host and thickness of particulate films. With the addition of P2VP, P4VP and amount of silver, morphology of the silver particulate films on PS/P2VP (50:50) and PS/P4VP (50:50, 75:25) could be modified to give the desired electrical results.

Author details

Pratima Parashar
Department of Materials Science, Mangalore University, Mangalagangotri, India
CET, IILM Academy of Higher Learning, Greater Noida, India

Acknowledgement

The author thanks DST for the XRD and NCL (Pune) for SEM facility. The author thanks DST for the funding through Women Scientist Scheme (WOS).

7. References

[1] Skofronick J G, Phillips W B (1967) Morphological Changes in Discontinuous Gold Films following Deposition. J. Appl. Phys. 38(12): 4791-4796.
[2] Fehlner F P (1967) Behavior of Ultrathin Zirconium Films upon Exposure to Oxygen. J. Appl. Phys. 38: 2223- 31.

[3] Kovacs G J, Vincent P S (1982) Formation and Thermodynamic Stability of a Novel Class of Useful Materials: Close-Packed Monolayers of Submicron Monodisperse Spheres Just below a Polymer Surface. J. Colloid. Interface. *Sci*. 90: 335-342.

[4] Kovacs G J, Vincentt P S (1983) Subsurface particulate film formation in softenable substrates present status and possible new applications. Thin Solid Films 100: 341-353.

[5] Kovacs G J, Vincentt P S, Trumblay C, Pundsak A L (1983) Vacuum deposition onto softenable substrates formation of novel subsurface structures.Thin Solid Films 101: 21-40.

[6] Kovacs G J, Vincentt P S (1984) Subsurface particulate film formation in softenable substrates present status and possible new applications.Thin Solid Films 111: 65-81.

[7] Kunz M S, Shull K R Kellock A J (1992) Morphologies of Discontinuous Gold Films on Amorphous Polymer Substrates. J. Appl. Phys. 72: 4458-4460.

[8] Mohan R K, Pattabi M (2001) Effect of polymer-metal particle interaction on the structure of particulate silver films formed on softened polymer substrates. J. New Mat. Electrochem. Systems 4: 11-15.

[9] Yuan JJ, Ma R, Gao Q, Wang YF, Cheng SY, Feng LX, Fan Z Q, Jiang L (2003) Synthesis and characterisation of Polystyrene/ Poly(4-vinylpyridine) Triblock Copolymers by Reversible Addition –Fragmentation Chain Transfer Polymerisation and Their Self-Assembled Aggregates in Water. J Appl Polymer Sci, 89:1017-1025.

[10] Bouslah N, Amrani F (2007) Miscibility and specific interactions in blends of poly [(styrene)-co-(cinnamic acid)] with poly (methyl methacrylate) and modified Poly(methyl methacrylate).Express polymer letters 1:44-50.

[11] Torrens F, Soria V, Codoner A, Abad C, Campos A (2006) Compatibility between polystyrene copolymers and polymers in solution via hydrogen bonding. Euro polymer Journal 42(10): 2807-2823.

[12] Bouslah N, Hammachin R, Amrani F (1999) Study of the compatibility of poly[styrene-co-(cinnamic acid)]/poly[(ethyl methacrylate)-co-(2-dimethylaminoethyl methacrylate)] blends. Macromolecular Chemistry and Physics 200(4):678-682.

[13] Jiao H, Goh S H, Valiyaveettil S (2001) Mesomorphic Interpolymer complexes and blends based on poly (4-vinylpyridine)- dodecylbenzenesulfonic acid complex and poly(acrylic acid) or poly(p-vinvylphenol. Macromolecules, 34:7162-7165.

[14] Kosonen H, Valkama S, Hartikainen J, Eerikainen H, Torkkel M, Jokela K, Serimaa R, Sundholm F, Brinke G, Ikkala O (2002) Mesomorphic Structure of Poly(styrene)-block-poly(4-vinylpyridine) with Oligo(ethylene oxide)sulfonic Acid Side Chains as a Model for Molecularly Reinforced Polymer Electrolyte. Macromolecules 35: 10149-10154.

[15] Wang X, Zuo J, Keil P, Grundmeier G (2007) Comparing the growth of PVD silver nanoparticles on ultra thin fluorocarbon plasma polymer films and self-assenbled fluoroalkyl silane monolayers. Nanotecnology 18:265303-265313.

[16] Heilmann A (2002) Polymer Films with Embedded Metal Nanoparticles, Berlin:Springler.

[17] Mayer A B R (2001) Colloidal metal nanoparticles dispersed in amphiphilic polymers. Polym. Adv. Technol. 12: 96-106.

[18] Beecroft L L (1997) Nanocomposite materials for optical applications. Chem.Mater. 9:1302-1317.

[19] Caseri W (2000) Nanocomposites of polymers and metals or semiconductors: Historical background and optical Properties Macromol.Rapid Commun. 21:705-722.

[20] Eilers H, Biswas A, Pounds T D, Norton M G (2006) Teflon AF/Ag nanocomposites with tailored optical properties. J.Mater. Res 21:2168-2171.

[21] Yonzon C R, Stuart D A, Zhang X, Mcfarland A D, Haynes CL, Duyne R P V (2005) Towards advanced chemical and biological nanosensors—An overview. Talanta 67:438-448.

[22] Saito R, Okamura S, Ishizu K (1992) Introduction of colloidal silver into a poly (2-vinyl pyridine) microdomain of microphase separated poly(styrene-b-2-vinyl pyridine) film. Polymer 33(5):1099-1101.

[23] Tao L, Ho R M, Ho J C (2009) Phase Behavior in Self-assembly of Inorganic/Poly(4-vinylpyridine)-b- Poly (ε-caprolactone) HybridMacromolecules 42:742-751.

[24] Pennelli G (2006) Lateral reduction of random percolative networks formed by nanocrystals: Possibilities for a new concept electronic device. Appl. Phys. Lett. 89:163513-163516.

[25] Rao K M, Pattabi M, Mayya K S, Sainkar S R, Murali Sastry M S (1997) Preparation and characterization of silver particulate films on softened polystyrene substrates. Thin Solid Films 310: 97

[26] Hassell T, Yoda S, Howdle S M., Brown P D (2006) Microstructural charecterisation of silver/polymer nanocomposites prepared using supercritical cabondioxide. J. of Phys: Conferrence Series 26:276-279.

[27] Dondos A S, Kondras P, Pierri P, Benoit H (1983) Hydrodynamic crossover in two-polymer mixtures from viscosity measurement. Macromolek Chem, 184(10):2153-2158

[28] Rao V, Ashokan P, Shridhar M H (1999) Studies on the compatibility and specific interaction in cellulose acetate hydrogen phthalate (CAP) and poly methyl methacrylate (PMMA) blend. Polymer 40:7167-7171.

[29] Shih K S, Beatty C L (1990) Blends of polycarbonate and poly(hexamethylene sebacate): IV. Polymer blend intrinsic viscosity behavior and its relationship to solid-state blend compatibility. Br Polym. J. 22 (1):11-17.

[30] Krigbaum W R, Wall F T (1950) Viscosities of binary polymeric mixtures. J Polym Sci. 5:505-514.

[31] Fox, T G (1956) Influence of Diluent and of Copolymer Composition on the Glass Temperature of a Polymer System..Bull Am Phys Soc. 1:123-125.

[32] Gordon M, Taylor J S (1952) Ideal copolymers and the second-order transitions of synthetic rubbers. I. Non-crystalline Copolymers. J Appl Chem 2:493-500

[33] Dong H.K, Won H.J, Sang C L, Ho C.K (1998) The compatibilizing effect of poly(styrene-co-4-vinylpyridine) copolymers on the polystyrene–polyethylene-based ionomer blends. J Appl. Polymer Sci. 69:807- 816.

[34] Parashar P, Pattabi M, Gurumurthy S C (2009) Electrical behaviour of discontinuous silver films deposited on softened polystyrene and poly (4-vinylpyridine) blends. J.Mater.Sci:Mater Electron 20: 1182-1185.

[35] Parashar P, Ramakrishna K, Ramaprasad A T (2011) A Study on Compatibility of Polymer Blends of Polystyrene/Poly(4-vinylpyridine). Journal of Appl Poly Sci 120 (3):1729-1735.

[36] Pattabi M, Rao K M, Sainker S R, Sastry M (1999) Structural studies on silver cluster films deposited on softened PVP substrate. Thin Solid Films 338:40-45.

[37] Rao K M, Pattabi M., Sainkar S. R, Lobo A., Kulkarni S K, Uchil J, Sastry M. S (1999) Preparation and characterisation of silver particulate structure deposited on softened poly(4-vinylpyridine) substrate. J. Phys. D: Appl.Phys. 32:2327-2336.

[38] Haoyan W, Eilers H (2008) Electrical Conductivity of Thin-Film Composites Containing Silver nanoparticles Embedded in a Dielectric Teflon AF Matrix. Thin Solid Films, 517: 575-581.

[39] Pattabi M , Rao K M (1998) Electrical behaviour of discontinuous silver films deposited on softened polyvinyl pyridine Substrate. J. Phys. D: Appl. Phys. 31: 19-23.

[40] Kiesow A, Morris J E, Radehaus C, Heilmann A (2003) Switching behavior of plasma polymer films containing silver nanoparticles. J.Appl. Phy. 94 (10): 6988-6990.

[41] Heilmann A, Kiesov A,Gruner M, Kreibig U (1999) Optical and electrical properties of embedded silver nanoparticles at low temperatures.Thin Solid Films 343-344:175-178.

[42] Fritzsche W, Porwol H, Wiegand A, Boronmann, Khler J M (1998) In-situ formation of Ag-containing nanoparticles in thin polymer film. Nanostrutured Materials, 10 (1) 89-97.

[43] Akamatsu K, Takei S, Mizuhata M, Kajinami A, Deki S, Fujii M, Hayashi S, Yamamoto K (2000) Preparation and characterization of polymer thin films containing silver and silver sulfide nanoparticles.Thin Solid Films 359:55-60.

[44] Carotenuto G (2001) Synthesis and characterization of poly (N-vinylpyrrolidone) filled by monodispersed silver clusters with controlled size. Appl. Organometal. Chem. 15(5):344-351.

[45] Kim J Y, Shin D H, Ihn K J, Suh K D (2003) Amphiphilic Polyurethane-co-polystyrene Network Films Containing Silver Nanoparticles. J Ind. Eng. Chem. 9(1):37-44.

[46] Heilmann A, Quinten M, Werner J (1998) Optical response of thin plasma-polymer films with non-spherical silver nanoparticles Eur.Phys. J.B, 3:455-461.

[47] Mandal S, Arumgam S K, Pasricha R, Sastry M (2005) Silver nanoparticles of variable morphology synthesized in aqueous foams as novel templates. Bull. Mater. Sci. 28(5):503-510.

[48] Kim JY, Shin DH, Ihn KJ (2005) Synthesis of CdS nanoparticles dispersed within amphiphilic poly(urethane acrylate-co-styrene) films. J.Appl.Polym Sci. 97(6):2357-2363.

[49] Biswas A, Bayer I S, Marken B, Pounds D, Norton M G (2007) Networks of ultra-fine Ag nanocrystals in a Teflon AF® matrix by vapour phase e-beam-assisted deposition Nanotecnology 18:305602-305608.

[50] Parashar P (2011) Morphology of Silver Particulate Films Deposited on Softened Polymer Blends of Polystyrene and Poly (4-vinylpyridine)J.Appl.Polymer Sci. 121 (2): 839-845.

[51] Takele H, Greve H, Pochstein, Zaporojtchenko V, Faupel F (2006) Plasmonic properties of Ag nanoclusters in various polymer matrices. Nanotechnology 17: 3499-3505.

[52] Parashar P (2011) Electrical behaviour of discontinuous silver films deposited on compatible Polystyrene/Poly (2- vinylpyridine) composite. J.Mater.Sci:Mater Electron, DOI 10.1007/s10854-011-0418-6.

[53] Parashar P (2011) Structural properties of silver particulate films deposited on softened polymer blends of polystyrene/poly (2-vinyl pyridine) J.Mater.Sci:Mater Electron, DOI 10.1007/s10854-011-0567-7

Chapter 2

Thermal Analysis of Phase Transitions and Crystallization in Polymeric Fibers

W. Steinmann, S. Walter, M. Beckers, G. Seide and T. Gries

Additional information is available at the end of the chapter

1. Introduction

Each year about 50 Million tons polymer is processed to fibers worldwide [1]. Polymeric fibers are manufactured into all sorts of daily as well as industrial goods [2, 3]. The most prominent materials are thermoplastic among which poly(ethylene terephtalat) (PET), polyamides (PA) and polypropylene (PP) make up the largest fraction [4]. Other thermoplastic polymers such as poly(vinylidene fluoride) (PVDF) belong to niche markets with highly specialized applications [5-7].

The most distinctive property of synthetic fiber materials which separates them from other polymeric products is the strong anisotropic material structure. Geometrically this is characterized by a rather high aspect ratio of diameter to length which can reach several magnitudes of order in filament fibers. An exemplary PET textile multi-filament bundle of 300 single filaments in one flat yarn weighs around 100 g per 10.000 meters length (100 dtex). This yields in a single filament diameter of roughly 3 μm. A common industrial bobbin holds up to 25 kg of a virtually endless length of yarn, which in this case would be 2.5 million meters. The predominant cross section is circular in shape. Nonetheless, depending on the fiber application other cross-sections are possible and also common [3,4].

On a structural level the strong anisotropic character of a thermoplastic fiber is mainly caused and influenced by the production process. Hence, the spinning process of thermoplastic fibers shall be explained briefly. Next to the direct spinning process in which the polymer is directly processed to fibers right after the polymerization process, the most common process is the extruder based fiber production. The polymer granules are heated and transferred into a molten state inside the extruder [8]. The melt is then conveyed into a gear pump which ensures a constant flow of mass. This constant polymer flow then is being pressed through filtration layers and finally extruded through capillaries. Following the extrusion the polymer is drawn down vertically and solidifies while cooling from extrusion temperature down to the ambient air temperature. Usually the fibers are drawn down mechanically by rollers [4,8]. Figure 1 illustrates a classical melt spinning line. The polymer granules are feed through a hopper (A) into an extruder (B). The molten polymer is transported through heated pipes (C) to a gear pump (D). The gear pump feeds the spin pack (E) in which several layers of filtration are placed. The polymer is then extruded through capillaries and exits the spinpack into the quenching zone (F) where a laminar air flow ensures constant cooling conditions. After solidification and before touching the first roller or godet (H) the filaments are usually coated with a spin finish (G).

The most important process parameters of the melt spinning process are: the extrusion temperature $T_{Extrusion}$; the mass flow through each single capillary $m_{Throughput}$; the density of the melt ρ_{Melt}; the cross-sectional area of the capillary $A_{Capillary}$; the viscosity η of the melt at the local temperature and the draw down speed v. Of course, there are numerous other parameters that affect the process such as the surface quality of the capillary walls, the form of the capillary rim, the ambient air profile consisting of flow direction, speed and temperature and others. For a basic comprehension these are neglected at this point. There is large number of extensive publications on these aspects available such as [9-13].

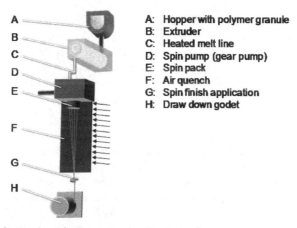

A: Hopper with polymer granule
B: Extruder
C: Heated melt line
D: Spin pump (gear pump)
E: Spin pack
F: Air quench
G: Spin finish application
H: Draw down godet

Figure 1. Schematic overview of a conventional melt spinning line

The cooling of the material from melt to ambient temperature takes place under tremendous stretching stress, which is characterized by the ratio between draw down and melt extrusion speed. This is commonly referred to as melt draw ratio (MDR). This melt draw ratio can vary between small one digit figures for rather thick filaments, e.g. fishing line applications, up to values well beyond 100 for fine filaments with diameters in the range of 1 to 50 μm. In most applications the fibers undergo a consecutive stretching or drawing stage after solidification. Thus this is called solid-state-drawing (SSD). Herein the filament is usually run between two rollers whereas the second roller is run at a higher speed than the first one. The speed ratio of the two rollers is referred to as the solid state draw ratio (SSDR) [2,11,12]. Usually, the solid-state-drawing is usually performed under elevated temperature levels.

This is usually realized by heated rollers, so that the fibers heat up before and after drawing when in contact with the surface of the rollers. Other principles can facilitate chamber ovens, contact heating plates or overheated steam. The process parameters of the drawing state also have a significant influence on the material structure, the orientation state and also the relaxation state. For example do fibers which are drawn and not properly heat-set a high degree of shrinkage which is usually unacceptable for most applications [2,4].

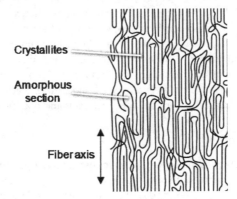

Crystallites

Amorphous
section

Fiber axis

Figure 2. Model for the morphology in a polymeric fiber in allusion to [14]

In both cases the melt drawing as well as the solid state drawing the macromolecules are oriented along the fiber axis (see figure 2). This results in a unique morphology and crystalline structure which can be found only in fibrous materials in this highly oriented state [9,11]. Although there remains some controversy and vivid discussion about the resulting crystalline structure, scientist and researchers agree in the fact that the highly elongated material exhibits deformed crystallites as well as a highly ordered amorphous phase in the non crystalline regions.

For the description of the crystalline structure, most commonly the Stacked-Lamellae and the Shih-Kebab models are quoted [14]. This is usually based on small angle x-ray scattering data, which however do not provide a real image of the structure [15]. A closed theory about the development of the various morphologies is not available. For the thermo-dynamical description of the crystallization the Gibbs free energy is used [16]:

$$\Delta G_m = \Delta H - T \cdot \Delta S \tag{1}$$

Elongated states cannot be discussed within the equilibrium thermodynamics, because of the missing isotropic character of the morphology. Nonetheless, an adoption of the equilibrium theories through consideration of anisotropic influences in analogy to the electro-magnetic field theory is possible [16,17]. In literature [16] within the discussion of entropy elasticity it is described that elongation of a melt through an outside force results in an increase of the free energy of the material. Thus a deformation will result in an increased Gibbs energy ΔG_m in spinning and thus will significantly influence the crystallization of the polymer [16].

For the ideal Gaussian polymer chain this energy contribution corresponds with a reduction of entropy $\delta S_{\Delta L}$ in the elongated state [11]:

$$\Delta G_{m,\Delta L} = \Delta H - T \cdot (\Delta S - \delta S_{\Delta L}) \qquad (2)$$

This change in entropy causes an increase of the crystallization temperature as well as an enhancement of crystallization rates. For polymorphic materials simple extensions of this theory are given in [18,19]. The different morphological phases are described through varying energy levels which depend on temperature and state of elongation and strain [19,20]. Depending on the process parameters and the material properties these phenomena are more or less prominent and detectable. Since the degree of crystallinity, the degree of orientation within the molecular structure and tendency of a material to crystallize when stored at temperatures above the glass transition temperature, thermal analysis is one key analytical method to investigate fiber materials, processes and fiber product properties.

2. Experimental method

The following text deals with the experimental investigation of polymeric materials. In this context the method of differential scanning calorimetry (DSC) is described and it will be pointed out what the general procedure is like and which experimental parameters have to be considered.

Differential scanning calorimetry follows the principle of the measurement of heat flow differences. By performing DSC a sample whose temperature is increased gradually and then subsequently cooled down is investigated and finally compared to a reference probe. Therefore it is possible to determine enthalpies and melting points of an arbitrary polymeric fiber. In this context a variation of the involved parameters offers a possibility to draw conclusions about underlying properties such as equilibrium values but concerning the execution of the experiment all of the possibly modifiable parameters have to be regarded carefully to perform DSC correctly. In order to perform DSC a furnace which can be heated up and cooled down homogenously is required. Inside this oven there are two mountings for the samples and each mounting is equipped with a high-sensitive temperature sensor [21,23,24]. The general set up is depicted in figure 3.

One mounting (left mounting in figure 3) is for the crucible which contains the prepared sample. The lid of the crucible has at least one hole to allow an exchange with the surrounding atmosphere. Furthermore, pressure build-up in the crucibles is prevented if parts of the sample vaporize.The other mounting (right mounting in figure 3) is for an empty crucible which functions as a reference. Due to the usage of such a reference only effects caused by the sample itself are observable in the final thermogram. The oven is purged with a gas (sample gas), so that transitions and chemical reactions in different atmospheres can be examined. To avoid oxidation processes a protective gas (e.g. N_2 or Ar) can be used to create an atmosphere around the sample during the process of DSC. Otherwise, air or oxygen can be selected. Furthermore, the space around the oven is purged with a protective gas (N_2) to avoid ice formation at low temperatures [21-24].

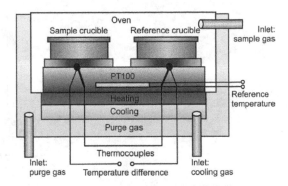

Figure 3. Schematic set up of differential scanning calorimetry

In a special form of differential scanning calorimetry a temperature-modulation is used. Usually one pre-defined frequency is used for temperature modulation [23,24]. The special TOPEM® (Mettler Toledo brand name) technique developed by Mettler Toledo allows the frequency-independent separation of reversing and non-reversing components of the heat flow by analyzing the impulse response of the sample to a pulse of stochastically varied length. Therefore a separation of overlapping effects is possible. Due to this separation extra insight can be gained [22].

3. Sample preparation

The process of sample preparation is of significant importance for the success of experimental investigation of polymeric fibers and has to be handled with great care. Additionally, different aspects of preparation have to be considered simultaneously in order to provide valid and reliable experimental results.

Before beginning with the experimental procedure itself, several aspects have to be dealt with. In general, both granules and fibers are treated in the same manner: The samples have to be reduced to small pieces so that they fit into the crucibles. In this context it is inevitably necessary to consider that the preparation method directly influences the results which are provided by differential scanning calorimetry. Therefore optimum conditions and parameters have to be found in order to determine certain effects (such as glass transition, crystallization or melting). Otherwise these effects still would be observable but not as good as if the optimum conditions were adjusted. Figure 4 conveys an idea which steps are necessary for sample preparation.

The sample has to be reduced to small pieces in order to perform the experiments. Therefore it is possible to alter the form and size of these pieces which yields different results. For example, the reduction to smaller pieces results in different observations concerning the gained thermogram. Usage of a sample with reduced size leads to a decrease of the peak and a lowered melting point. Therefore one can conclude that the mechanical aspect of preparation cannot be neglected and has to be treated carefully. Figures 5 a) and b) deliver an impression which mechanical appearance of the sample has to be chosen in the best case.

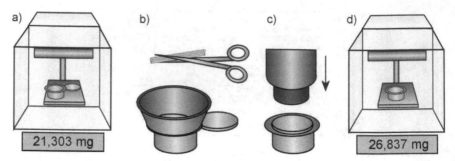

Figure 4. Preparation of fiber samples for DSC:
a) Determination of the empty weight of the crucible
b) Reduction of the fiber to small pieces
c) Insertion of the crucible into cavity
d) Determination of the total weight.

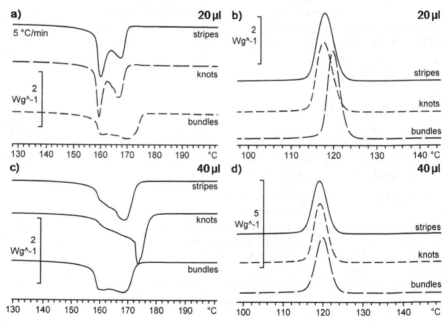

Figure 5. Effect of preparation conditions and crucible size
a) Heating process with crucible volume of 20 µl,
b) Cooling process with crucible volume of 20 µl,
c) Heating process with crucible volume of 40 µl,
d) Cooling process with crucible volume of 40 µl.

Obviously, best results will be achieved if the sample is manufactured into short fiber chops (stripes). Especially knots cause irregularities as an unsteady graph in the thermogram. The explanation for this phenomenon is that stripes offer an increased contact surface whereas knots melt discontinuously. Additionally, this result is independent on the volume of the crucible. In figures 5 c) and d) a similar thermogram is presented but the volume of the crucible is doubled (from 20 μL to 40 μL) and what is apparent here is the fact that again the short fiber chops provide the steadiest graph. Additionally, one can conclude that only the usage of the crucible with 20 μL volume delivers reliable results due to the fact that for the 40 μL crucible the graph inside the thermogram is rather uneven. This effect is caused by the better heat contact in the smaller crucibles, since the lid is pressed to the bottom of the crucible during the preparation process.

Concerning the weighted portion another remarkable effect can be observed: The peak height increases whereas the peak width drastically increases (see figure 6). Furthermore the onset temperature also increases logarithmically as a function of the sample mass. Nevertheless this result is rather obvious due to the fact that for the melting process of a sample with increased weight a higher amount of energy is required than for a sample with less weight. Concerning the interpretation of the results an advantage of less sample weight is that the peak sharpness is increased and therefore overlapping effects can be observed easier. In this context it is necessary to keep in mind that it is possible that thermal events that cause only little effects might be missed. These observations are depicted in the following diagram (figure 6), where the double peak during melting of polypropylene gets more separated for lower sample masses.

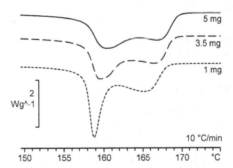

Figure 6. Normalized thermograms for different sample weight of polypropylene fibers

4. Influence of experimental parameters

To illustrate the effects caused by the variation of different parameters, several thermograms are shown in the following paragraph. All these thermograms refer to the investigation of fibers and as an example polypropylene (PP) was used. Firstly, the heating rated is varied as it is depicted in figure 7 a).

Analysis of the thermogram presented above conveys that the alteration of the heating rate causes a strong effect on the results delivered by differential scanning calorimetry. With all other parameters remaining constant the peak height and width increases with increasing heating rate. Moreover, a double peak is of importance. Phase transitions only occur or can be separated from the melting process if lower heating rates are used. For high heating rates they cannot be observed.

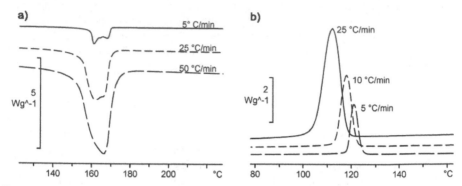

Figure 7. a) normalized thermograms for different heating rates for polypropylene fibers b) normalized thermograms for different cooling rates of polypropylene granule

Secondly, the variation of the cooling rate is also important as it is depicted in figure 7 b). Beside other observations it is very obvious that the crystallization peak is shifted in positive temperature direction with decreasing cooling rates and the absolute height of the peak is also decreasing. Apparently the correct choice of the cooling rate is as important as it is for the heating rate for the analysis of crystallization processes. By taking into account different cooling rates and extrapolating the peak temperature to isothermal conditions (not measurable in DSC), the true crystallization point can be evaluated.

4.1. Fibers from commodity polymers

All results provided by the last paragraph yield that the alteration of heating and cooling rate causes a strong effect on the results of DSC. Due to this fact it is vital for the success of the analysis to consider the effect of the altering of especially those parameters mentioned above on the properties of a polymeric material which are in the center of interest. Among others the properties crystallization, melting and glass transition shall usually be investigated. In order to support the successful performance of DSC several hints and recommendations for various polymers will be presented in this paragraph. Typical commodity polymers are polypropylene (PP) [25], polyamide (PA6) [26] and polyethylene terephthalate (PET) [27]. In the following paragraph the effects of variations of heating and cooling rate on these polymers will be examined and presented.

Starting with PP (LyondellBasell Moplen HP561R [25]), in the following thermogram the results of DSC with different heating rates are depicted (figure 8 a)).

With increasing heating rate an increase of the peak height and width is noticeable. Additionally, the experiment's velocity is decreasing and the melting process starts earlier.

Noticeable is also the double peak when low heating rates were used. During the melting process of fibers it is likely that a phase transition from α- to γ-phase takes place [28]. The obversation of this transition is depending on the choice of the heating rate because it is possible that the material melts directly or the effect is superimposed by others. Due to the fact that a characteristic amount of energy is necessary, it is possible to observe this transition as a peak in the thermogram if the heating rate is chosen correctly. After complete transition to γ-phase the sample will melt completely and another peak is observable in the thermogram. Now the distance between these mentioned peaks is depending on the heating rate and if this rate is adjusted inappropriately both effects are no longer separate from each other. Another possibility is that no phase transition occurs because the temperature is risen quickly enough to start the melting process directly.

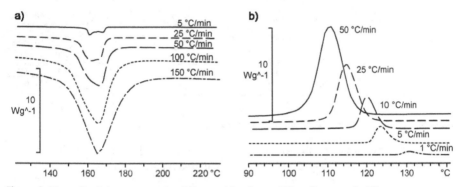

Figure 8. Normalized thermograms for different a) heating and b) cooling rates for PP

In the thermogram (figure 8 b)) the variation of the cooling rate is presented for PP. With increasing cooling rates the crystallization peak and its width increases. As reasonably expected the crystallization procedure is slower for higher cooling rates. The actual crystallization peak is determined by extrapolating a virtual crystallization peak for a heating rate of 0 °C/min from the peaks for known heating rates.

As the previous results were gained from investigations of PP similar experiments were performed for PA6 (BASF Ultramid B24N03 [26]). Beginning with a variation of the heating rate as it is depicted in figure 9 a).

Similar to PP the peak width and height is increased for higher heating rates. In this case another remarkable effect can be observed:

With heating rates higher than 25 °C/min a glass transition is starting below 50°C. This indicates that if heating rates below 25 °C/min were chosen this kinetic transition causes a too small effect and is therefore not visible. If the heating rate is then increased further another peak becomes observable. What is visible here is a phase transition from γ- to α-phase between 120 °C and 160°C [29]. For the observation of these two effects, it is therefore necessary to choose a heating rate above 25 °C/min.

In analogy the cooling rate is varied as it is depicted in figure 9 b). Similar to PP, the crystallization peak shifts to lower temperatures for higher heating rates and true crystallization point can be gained by an extrapolation of the peak temperature. The glass transition point cannot be observed since the actual state of the polymer chains is frozen in during cooling and changes in heat capacity will only be visible by heating up the sample again. Furthermore, the phase transition is also not visible since the material crystallizes in the α phase.

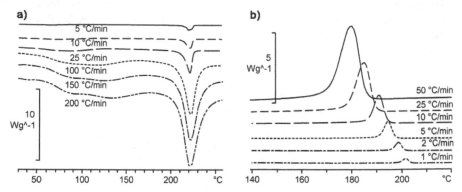

Figure 9. Normalized thermograms for different a) heating and b) cooling rates for PA6

Finally, DSC is performed for PET (Invista Polyester Chips 4048 [27]) and again heating and cooling rates are varied. Firstly, the heating rate is altered (figure 10 a)).

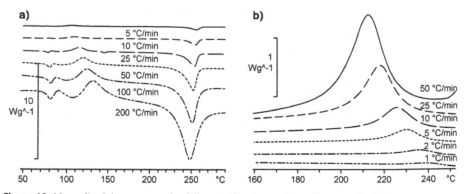

Figure 10. Normalized thermograms for different a) heating and b) cooling rates for PET

Again, the peak width and height increases with higher heating rates. But the glass transition is in this case even more remarkable: Beginning with a heating rate of 25 °C/min this transition becomes visible and is much more pronounced than it was in the investigation of PA6, since the mobilization of polymer chains during the transition has a larger influence on the heat capacity. Furthermore, a relaxation process can be observed shortly after glass transition, indicating that the elongated state of the amorphous polymers chains was frozen in during the rapid cooling in the spinning process. At high heating rates another effect can be observed: after the glass transition and relaxation second crystallization occurs, which pronounces the quenching of the material during the spinning process. After heating over the glass transition point, the mobility of the amorphous or mesomorphous polymer chains increase so that the already present crystallites can grow.

Secondly, the cooling rate is changed for PET sample fibers (figure 10 b)). The analysis of this diagram yields similar results as for the investigation of PP and PA6. With increased cooling rates higher and broader crystallization peaks become visible. Additionally, these peaks are shifted to lower temperatures with increased cooling rates.

4.2. Recommendations

Using the results from the previous paragraph, one can conclude that the appropriate choice of heating and cooling rates is essential for the experiment's success. Therefore, the following table contains valuable information concerning the right choice of these rates in order to gain maximum benefit from performing differential scanning calorimetry of commodity polymers. In this context it is necessary to consider that with too high heating rates some effects may superimpose and therefore might not be visible.

Polymer	Process	Effect	Temperatures	Rates
PP	Heating	Phase transition ($\alpha \rightarrow \gamma$)	156 – 162 °C	< 25 °C/min
PP	Heating	Melting	162 – 170 °C	all
PP	Cooling	Crystallization	100 – 125 °C	all
PA6	Heating	Glass transition	50 – 60 °C	> 25 °C/min
PA6	Heating	Phase transition ($\gamma \rightarrow \alpha$)	100 – 170 °C	> 10 °C/min
PA6	Heating	Melting	200 – 240 °C	all
PA6	Cooling	Crystallization	160 – 200 °C	all
PET	Heating	Glass transition with relaxation	70 – 90 °C	> 25 °C/min
PET	Heating	Crystallization	90 – 160 °C	> 10 °C/min
PET	Heating	Melting	220 – 270 °C	all
PET	Cooling	Crystallization	190 °C – 240 °C	all

Table 1. Recommendations for the right choice of heating and cooling rates

5. Poly(vinylidene fluoride) fibers

Poly(vinylidene fluoride) is a thermoplastic, semicrystalline fluoropolymer with the monomer unit [CF_2-CH_2]. Due to the high fluorine content, it exhibits excellent chemical stability [30]. Furthermore, the polar side groups are responsible for the piezoelectric, pyroelectric and ferroelectric properties of the material, which are only present in one crystalline phase of the polymorphic material, the so called β phase [31,32]. In total, four different crystalline phases can occur [33]. An overview of the crystalline phases, together with the conditions for their formation, is given in table 2.

Crystalline phase	Molecular conformation	Crystalline unit cell (a-b-projection)	Polarity	Conditions for formation
α	TGTG′		non-polar	Cooled from melt, cast from solution
β	TT		polar (high)	Mechanical stress, high electric field
γ	TTTGTTTG′		non-polar	Heat treatment, cast from solution
δ	TGTG′		polar (low)	Electric field

Table 2. Crystalline phases of poly(vinylidene fluoride) [34-37]

If the material is present in the β phase, it can be used to create sensors or actuators, which are commonly used in the form of films in microphones, hydrophones or headphones. Here, the necessary process conditions (drawing of the films or high electric fields) for the β phase formation are well known [31,38].

In the case of fibers, the material could potentially be used as sensor or actuators. Possible applications include direction sensitive and spatially resolved strain measurement, which is useful for health monitoring in medical / smart textiles or structural health monitoring in fiber reinforced composites. However, suitable process conditions for fiber spinning, drawing and further processing steps have to be found, whereas β phase crystallites have to be formed and not be destroyed along the process chain [39-42]. Therefore, methods of thermal analysis were developed to identify the presence of the β phase, which are validated by additional X-ray diffraction measurements (WAXD) and dynamic mechanical analysis (DMA) [40,42]. In the following section of this chapter, these methods and their validation will be demonstrated, and they will be applied to gather information about phase transitions during melt spinning, drawing and heat treatment. Therefore, thermal analysis is a powerful tool for process analysis and the development of a process chain for the creation of piezoelectric sensor fibers.

6. Experimental details

Conventional and temperature modulated DSC are carried out on a DSC 1 from Mettler Toledo, Greifensee, Switzerland, equipped with a FRS5 sensor having 56 thermocouples. During the experiments, heating and cooling rates were varied between 1 °C/min and 20 °C/min in a temperature range between -90 °C and 250 °C, which corresponds to 50 °C below the glass transition and 70 °C above the melting point. As checked by thermogravimetric analysis, no weight loss and therefore no polymer degradation occurs in this temperature region during the relevant residence times. For the temperature modulated DSC analysis, TOPEM® technique by Mettler Toledo was used, where the constant heating rate was modulated with heat pulses of 0.5 °C height and stochastically varied length between 15 and 30 s (corresponds to 33.3 and 16.6 MHz). Experimental conditions for DSC analysis are summarized in table 3.

Parameter	Value	Unit
Starting temperature	-90	°C
End temperature	250	°C
Heating / cooling rate	1 / 2 / 5 / 10 / 20	°C/min
Constant heating / cooling rate (TOPEM®)	0.5 / 1 / 2	°C/min
Heat pulse height (TOPEM®)	0.5	°C
Heat pulse length (TOPEM®)	15 - 30	s
Crucible size (granule)	40	µl
Crucible size (fiber)	20	µl
Purge gas volume rate	50	ml/min

Table 3. Parameters for DSC experiments

Results of thermal analysis are compared to the polymer structure and polymer chain orientation, which are determined by wide angle x-ray diffraction (WAXD). Here, a 2D image plate system IPDS II from STOE & Cie GmbH, Darmstadt, Germany, is used for simultaneous analysis of structure and texture. Since the three most important crystalline forms of PVDF α, β and γ have unique diffraction patterns, the method can be easily used to identify the phases. By additional experiments with a heating chamber and heating rates similar to DSC, the underlying phase transitions can be assigned directly to their thermal effect. Since β phase formation can be achieved by mechanical stress, thermal properties of the material are further correlated to dynamic mechanical analysis (DMA), which is carried out on a DMA/SDTA861e by Mettler Toledo. Here, mechanical relaxation processes determined from the phase shift tan(δ) between storage and loss modulus are correlated to their contributions to heat flow.

7. Properties of raw material

In this chapter, some information about the thermal properties and the crystallization behavior of the raw material (Solvay Solef® PVDF 1006 [43]) will be given, which can be later on compared to changes in the fiber material. A typical DSC thermogram (with heating and cooling rate of 10 °C/min) of PVDF quiescently cooled from the melt (with a cooling rate of 1 °C/min) can be found in figure 11. As indicated in the figure, the melting and crystallization peak can be identified clearly.

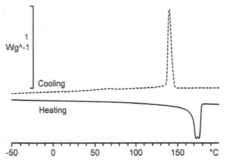

Figure 11. DSC thermogram (heating and cooling) for PVDF bulk material

The endothermal peak correlates to the melting of the material, which takes place in the temperature region between 165 and 180 °C. During the cooling phase, the crystallization (exothermal peak) takes place between 150 and 140 °C. When compared to X-ray data and polarizing microscopy, this behavior can be assigned to a non-textured α phase (figure 12 a), orientation factor f = 0) with spherulitic morphology (figure 12 b)). The material properties are summarized in table 4.

Figure 12. a) Diffraction pattern of quiescently cooled PVDF sample,
b) Polarizing microscopy image of quiescently cooled PVDF sample.

8. Influence of the spinning process

PVDF fibers (multifilaments) are produced in an industrially relevant high speed spinning process. The winding speed is varied between 100 and 2.500 m/min, which correlates to a melt draw ratio of 40 respectively 100. Due to this high draw ratios, polymer chains are oriented and orientation induced crystallization takes place. However, mechanical stress is relatively low during the melt drawing, so that the material crystallizes in a textured α phase (orientation factor f $\approx$ 1). This information can be extracted from the X-ray data (figure 13 a)).

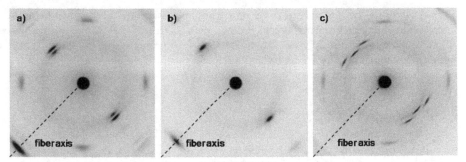

Figure 13. a) Diffraction pattern of melt spun PVDF fiber (winding speed 2.500 m/min), b) Diffraction pattern of a drawn PVDF fiber (DR = 1.5, drawing T_D = 140 °C), c) Diffraction pattern of a thermally treated PVDF fiber (T = 165 °C).

Depending on the heating rate, the melting behavior changes in the fiber material. After the melting peak, a second peak occurs (figure 14 a)). By further analysis (section "phase transitions during heat treatment"), this peak correlates to the melting of the γ phase, which is converted from α phase during the heating process. This peak gets larger for lower heating rates, since the process of γ phase conversion has more time to take place. Further evidence for changes in this kinetics can be found in the shape of the melting peak, whereas the first part of the peak becomes larger for lower heating rates. Furthermore, γ phase formation is more promoted in fibers produced with higher winding speeds (figure 14 b)). Here, also the same effect on the melting peak (shift to the first part) can be found.

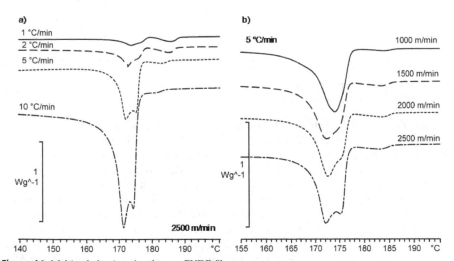

Figure 14. Melting behavior of melt spun PVDF fibers:
a) as a function of the heating rate,
b) as a function of the take up velocity.

Typical values for thermal and structural properties for a melt spun fiber can be found in table 4. Even though the material is cooled down with about 1.000.000 °C/min during fiber spinning, the melting enthalpy is exactly in the same range compared to the slowly cooled sample. This emphasizes the strong enhancement of crystallization rates due to the uniaxial orientation in the spinning process.

9. Influence of the drawing process

In the drawing process, a plastic deformation of the material in the solid state takes place, whereas the drawing ratio DR and drawing temperature T_D can be varied. Starting from the standard process (DR = 1.4, T_D = 140 °C), drawing ratio and temperature are varied separately between 1.0 and 1.6 as well as 40 °C and 160 °C respectively. The effect of both parameters on the melting behavior is displayed in figure 15.

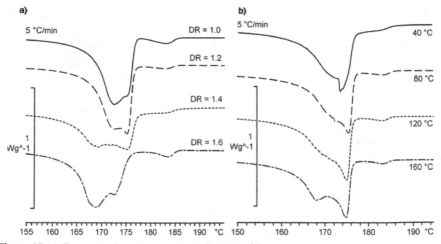

Figure 15. a) Changes in the melting peak as a function of draw ratio,
b) Changes in the melting peak as a function of drawing temperature.

A main effect of the drawing process is a modification of the melting peak, whereas peak temperature shifts to lower temperatures with higher drawing ratios (168 °C for the highest draw ratio compared to 173 °C for an as-spun fiber). In the X-ray diffraction pattern (figure 13 b)), the formation of β phase can be identified, so that the modification of the melting peak correlates to this crystalline phase. Like in the undrawn fiber, the peak coming from γ phase melting can also be found. The same effect can be found by increasing the drawing temperature. The underlying phase transitions during the melting process will be described in detail in the next section. At lower temperature regions, further peaks can be found (figure 16) depending on the drawing temperature. For low temperatures, a signature close after the glass transition point (-35 °C) can be found. A second peak can be found at higher temperatures, which is present in all fibers. For undrawn fibers, the peak temperature is around 55 °C and for drawn fibers about 5 °C higher.

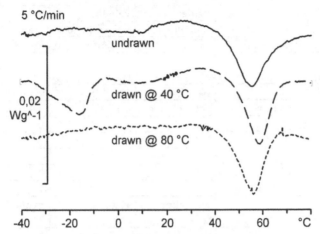

Figure 16. Relaxation processes at different drawing temperatures (draw ratio DR = 1.4)

Both peaks correlate a relaxation process, which can be found in DMA measurements, which are displayed in figure 17. Here, glass transitions can be identified easily. Another relaxation process (known as α_c relaxation in other types of polymers [44]) can be found at 55 °C (α phase) and at 60 °C (β phase), so there is clear evidence for a heat contribution to DSC measurements of this relaxation process. A reason for the occurrence of the first peak in coldly drawn fibers is a drawing temperature below the relaxation temperature, so that more energy is stored in the material and released when heated above glass transition temperature.

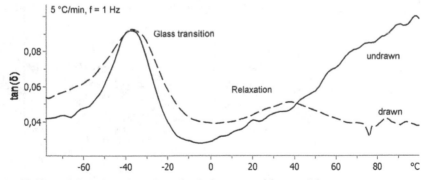

Figure 17. Phase shift tan(δ) in dynamic mechanical response of the material

When compared to X-ray diffraction measurements, β phase amount is drastically increased without affecting overall crystallinity if the fibers are drawn above the relaxation temperature. Therefore, the relaxation process takes place in the α phase and promotes β phase formation. The properties of all fibers are summarized in the table 4.

Parameter	Value	Value	Value	Unit
Type of material	Granules	Melt spun fiber	Drawn fiber	-
Relaxation temperature range	-	36 - 65	40 - 71	°C
Relaxation temperature peak	-	55	60	°C
Relaxation enthalpy	-	1.6	1.3	J/g
Melting range	168 - 180	168 - 185	165 - 185	°C
Melting peak temperature	174	173	168	°C
Melting enthalpy	49.6	48.2	45.8	J/g
Crystallization range	140 - 150	-*	-*	°C
Crystallization peak temperature	145	-*	-*	°C
Crystallization enthalpy	50.0	-*	-*	J/g
Crystalline phase	α	α	β	-
Morphology	spherulitic / non-textured	textured	textured	-

Table 4. Thermal and structural properties of PVDF (*: not detectable since determined by process)

10. Phase transitions during heat treatment

Heat treatment is important in the further processing of the fibers. If they are to be used as piezoelectric sensors, a polarization process has to take place at elevated temperatures without having a reconversion of β phase to α phase. To identify such a transition and explain phase transformations during the melting procedure of the material, temperature modulated measurements can be used. However, the occurrence of different crystalline phases has to be validated X-ray measurements of heated fibers. The results for temperature modulated measurements (TOPEM ®) are displayed in figure 18 for a undrawn and a highly drawn fiber, whereas heat flow is separated into reversing and non-reversing parts.

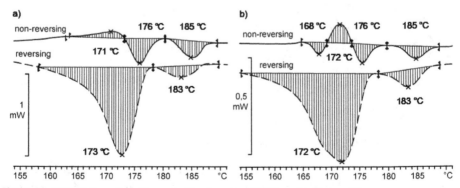

Figure 18. TOPEM ® results with constant heating rate of 2 °C/min (melting peak):
a) PVDF fiber containing α phase,
b) PVDF fiber containing β phase.

For an interpretation of the results, the highly oriented crystalline structures have to be taken into account. They were formed far from equilibrium and therefore must contribute to the non-reversing heat flow. On the other side, energy stored in the solidified amorphous parts due to short-range bonding to neighbor polymer chains contributes to the reversing heat flow. However, also crystalline phases contribute to the reversing heat flow if time scale of the phase transitions is larger than the pulse length applied to the sample. In total, 3 (α phase fiber) or 4 (β phase fiber) transformations in the crystalline regions take place. The same amount of phase transitions can also be found in X-ray measurements. The additional peak in the β phase fibers is caused by a reconversion to the α phase. After this transition, the changes in the crystalline structure are the same for both fibers, starting from α phase conversion to γ phase, going on with the melting of the remaining α crystallites and finally ending with the melting of the γ phase. Since the melting of the α phase takes place shortly after the conversion to γ phase, the amount of γ phase (compare to figure 14) strongly depends on the heating rate. All phase transitions during heat treatment are summarized in table 5.

Transition	Temperature range (peak)	Heat flow
$\beta \rightarrow \alpha$	165 °C – 170 °C (168 °C)	endothermal
$\alpha \rightarrow \gamma$	166 °C – 173 °C (171 °C)	exothermal
$\alpha \rightarrow$ melt	173 °C – 180 °C (176 °C)	endothermal
$\gamma \rightarrow$ melt	180 °C – 190 °C (185 °C)	endothermal

Table 5. Summary of phase transitions in PVDF fibers during heat treatment

11. Overview over phase transitions

For the selection of the right process settings for the production of piezoelectric fibers, it is necessary to get an overview of all phase transitions which can occur in the processes (melt spinning, solid state drawing and heat treatment/polarization). All these transitions were described in the previous sections and are summarized in figure 19.

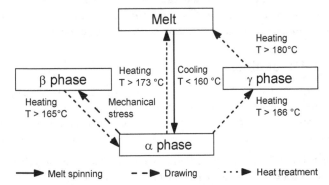

Figure 19. Overview of phase transitions in poly(vinylidene fluoride) fibers

After melt spinning, the fibers should be drawn at high ratios (close to maximum elongation) to form as much β phase fraction as possible. Drawing temperature should be at least above 55 °C to enhance β phase formation and prevent relaxation processes. For further polarization processes, temperature should be kept below 160 °C to keep the material in the β phase. For the process development, calorimetric measurements are of great value to identify the phase transitions, so that they can be realized in the process by choosing the right parameters. Furthermore, thermal analysis is of great value for further quality control, since the present crystalline phases can be detected by a simple method (compared to techniques like X-ray diffraction).

12. Carbon nanotube composite fibers

Carbon nanotubes (CNT) are allotropes of pure carbon in a form of a cylindrical structure. Depending on the number of graphite sheets forming the tube, they are divided into single wall (SW-CNT) and multi wall carbon nanotubes (MW-CNT). Beside excellent mechanical properties, CNT offer high electrical and thermal conductivity. When added to a polymer matrix, their properties are partially transferred to the polymer nanocomposite material [45]. In the case of mechanical reinforcement, an increase of Young's modulus can be observed in many polymer matrices [45,46]. One of the most interesting aspects is the formation of electrical conductive paths, so called percolative networks, in otherwise insulating materials [45,47]. Compared to other conductive fillers like carbon black (CB), the amount of CNT for reaching electrical conductivity is extremely low, whereas percolation thresholds below 0.1 % were observed [48].

However, not only the desired material properties in solid state change by the additivation of CNT. The nanoparticles interact with the polymer matrix and influence the rheological properties of the polymer melt, but also the crystallization behavior [46]. Especially in the melt spinning process, process settings have to be adjusted to allow the production of CNT modified nanocomposite fibers [49-53]. Therefore, the nancomposites have to be analyzed by DSC to understand the effects of CNT on the different polymer matrices, so that the right process parameters for fiber production can be found. Furthermore, thermal analysis provides useful information about the functional properties of the material, since crystallization conditions have a large influence on the electrical conductivity.

13. Experimental conditions

For each polymer, conventional DSC was carried out at least 50 °C below the glass transition (except of polyethylene, where glass transition is too low) and 50 °C above the melting point. Heating rate was varied between 2 °C and 20 °C, whereas results with 5 °C/min are shown in this chapter, since all effects to be demonstrated can be found at this heating rate.

Electrical properties of the composite materials where checked with a LCR meter, whereas the materials were tested in a frequency range between 1 Hz and 100 kHz. By checking AC conductivity, effects of partially not connected conductive networks can be found.

Furthermore, the values determined at low frequencies can be considered as the DC conductivity of the material.

14. Compounding in different polymer matrices

Depending on the type of polymer, CNT can influence the crystallization process in two different ways. Both effects will be described in this section with the help of the examples polypropylene (PP, Basell Moplen HP561R [25]) and polyamide 6 (PA6, BASF B24N03 [26]).

In PP (compare figure 20 a)), crystallization peak shifts to higher temperatures with increasing amount of CNT. This effect has two reasons. First, the high thermal conductivity of CNT allows the latent heat to be transferred faster from the polymer to the surrounding medium, to that sample temperature is closer to the reference temperature compared to unmodified samples. However, this effect shifts the peak only about 0.5 °C per w% CNT added to the polymer. The more dominant effect is the acting of CNT as foreign substance in the polymer, on which crystal nuclei can be formed at higher temperatures. After this nucleation, crystallites can grow in the usual way. By adding only 1 w% of CNT, this effect shifts the melting peak by 5 °C to higher temperatures.

The effect in PA6 is different (see figure 20 b)). The form of the crystallization peak changes to that a double peak can be found. Here, the part of the peak at lower temperatures (which is also shifted compared to unmodified material), has the same origin like in PP. Here also higher crystallization temperature is caused by enhanced thermal conductivity and crystal nucleation on the CNT. The part of the peak at higher temperatures is caused by another effect, the direct crystal growth on the CNT. The particles act as nuclei for the polymer chains and the crystallites can grow at higher temperatures, since no undercooling is needed for crystal growth compared to nucleation.

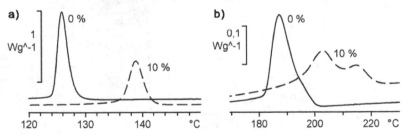

Figure 20. Effect of CNT on polymer crystallization (cooling rate 10 °C/min):
a) Polypropylene (PP),
b) Polyamide 6 (PA6).

The type of crystallization process has a large influence on the electrical properties of the material, especially on the percolation threshold. A comparison of the electrical properties of the two polymers can be found in figure 21 a). The percolation threshold is the value for CNT concentration, where the specific resistivity drastically decreases over several orders of magnitude. Above the threshold, resistivity only decreases slowly due to the higher amount of conductive filler. For polypropylene, the threshold can be found at 3 w% compared to 7 w% for polyamide 6.

For polypropylene, CNT act as nucleation seeds, but the material has no chemical affinity to the nanoparticles. Therefore CNT try to form aggregates in the polymer matrix, which can build conductive paths through the whole fiber. The region between the aggregates is then filled by PP crystallites during the crystallization process. Because of the separation of the different regions, only lower amount of CNT is needed to form conductive paths. Therefore, the dynamic percolation threshold defined by the crystallization process is significantly lower than it would be for a uniform distribution of the particles in the fiber. This behavior can be observed with transmission electron microscopy (figure 21 b)).

For polyamide 6, where polymer crystallites directly grow on the particles, a chemical affinity between the components is given and single CNT are separated by the polymer matrix (see figure 21 c)). Therefore, CNT are well distributed in the polymer and the percolation threshold is defined by the geometry of the nanoparticles (length and diameter) as well as their orientation to the fiber matrix. Single CNT in PA6 can be oriented better compared to the aggregates in PP, so that the resistivity of a fiber is highly sensitive to the process parameters in the spinning and drawing process.

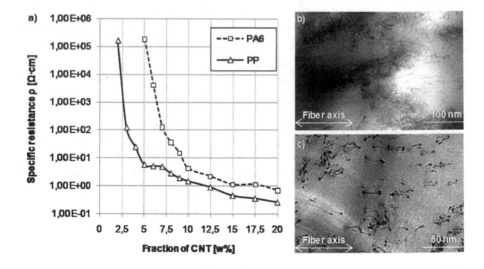

Figure 21. a) Electrical conductivity as a function of CNT weight fraction in different polymer matrices (PP and PA6),
b) Transmission electron microscopy image of 10 w% CNT in PP,
c) Transmission electron microscopy image of 5 w% CNT in PA6.

15. Influence of the spinning process

As described above, CNT have a major influence on the crystallization behavior in the different polymers. Even though the material is cooled down much faster during melt spinning and orientation induced crystallization is dominating the structure formation in the material, crystallization phenomena are drastically altered due to the presence of the nanoparticles. An example for the changes in the crystallization of polypropylene can be found in figure 22. If the material is heated above the relaxation point (approx. 50 °C), recrystallization phenomena can be found in the unmodified fibers. If CNT are doped to the PP matrix, no recrystallization peak can be found. Since the melting enthalpy is constant for different CNT fractions compared to the raw material (after recrystallization), it can be concluded that crystallization rates are enhanced by the presence of the particles. This observation is correlating with the fact, that the material cannot be drawn at high ratios in the molten state if CNT are present.

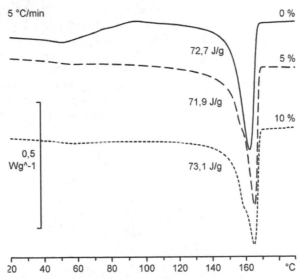

Figure 22. Changes in the melting behavior as a function of CNT concentration in polypropylene fibers spun with a winding speed of 50 m/min

In polyamide 6, the thermal property affected mostly is the glass transition. This effect is displayed in figure 23. By adding higher amounts of CNT, the glass transition temperature shifts to higher values. Thus a T_G of 60 °C is reached for a concentration of 10 w% compared to 52 °C for the same type of unmodified material. Since polymer crystallites grow on the CNT surface, the material is the polymer matrix has a better chemical bonding to the nanoparticles and the polymer chain mobility is lowered due to their presence. Therefore, more energy is needed to mobilize the polymer chains resulting in a higher glass transition temperature.

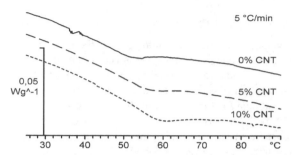

Figure 23. Changes in the glass transition as a function of CNT concentration in polyamide 6 fibers spun with a winding speed of 50 m/min

16. Effects in blend systems

In blend systems, two or more polymers are mixed together without forming a mutual polymer chain. If both components are not compatible with each other, they are separated in the fiber material, but influence each other [54]. Depending on the chemical structure of the polymers, CNT are attracted more by one of the components and aggregate in this material. This effect will be explained on a mixture of polypropylene (PP, LyondellBasell Moplen HP561R [25]) and low-density polyethylene (LDPE, LyondellBasell Lupolen 1800S [55]). The crystallization and melting behavior can be found in figure 24. In the crystallization process, the PP peak shifts by the additivation of CNT, which indicates the nucleation of PP

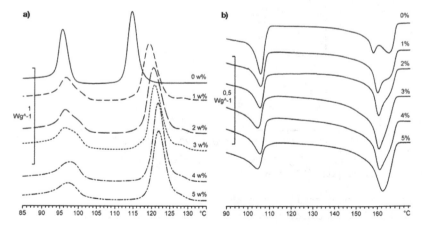

Figure 24. Effect in blend systems of polypropylene / polyethylene (1:1, heating/coolingrate 5 °C/min) as a function of CNT concentration:
a) Crystallization phenomena during cooling of bulk material,
b) Melting phenomena during heating of fibers.

on the CNT and therefore the presence of nanotubes in this polymer. For LDPE, there is no temperature shift of the crystallization peak. However, the peak gets broader. This indicates major changes in the phase separation behavior, but without changing the crystallization process by the nanotubes. Possible reasons for the broadening are more diverse sizes of regions of LDPE in PP, which tend to crystallize at different temperatures. Further evidence of the aggregation of CNT in PP can be found in the melting process. While the form of melting peak of LDPE is not changed at all, the PP changes from a double peak to a single peak. Since the occurrence of the double peak indicates a phase transformation before the actual melting procedure, it can be concluded that the present crystalline phase (α phase) is stabilized by the presence of CNT and the nanotubes are situated in the PP phase of the blend system.

17. Impact on the development of electrically conductive fibers

As stated above, the choice of the base polymer as well as the melt spinning of the fibers has a large influence on the electrical conductivity. The base materials can be classified into two groups. In polymers like polypropylene, CNT act as nucleation seeds. The particles can aggregate and easily form conductive paths, so that a low percolation threshold can be observed. This class of materials is very useful for the production of electrically conductive fibers with high spinning speeds, since electrical conductivity is less affected by the spinning process and possible mechanical treatment during the use of the fibers. For polymers like polyamide 6, crystallites grow on the surface of CNT, so that particles are separated and the percolation threshold is higher. Since single CNT can be oriented easily, the spinning and drawing process has a larger influence on the conductivity. This makes the processing of the materials more instable. However, they can potentially be used to create sensors, whereas mechanical stress and deformation can be detected by changes in the resistivity.

18. Conclusion

Taking into account the unique morphology of polymeric fibers, special analysis methods are needed for the development of new materials and processes. Due to high deformation of the molten polymer in the spinning process as well as the solid-state deformation during drawing, the polymer chains are highly oriented along the fiber axis. Differential scanning calorimetry, especially with temperature modulation, is one of the most important tools for fiber development. It can be used for a fundamental study of new materials and their behavior during spinning and processing as well as quality control of commidity polymers.

For the analysis of commodity polymers, the effect of sample preparation and experimental parameters is demonstrated. For the preparation of samples, small crucibles with fibers cut into small pieces are useful to measure thermograms with clearly visible effects. The choice of parameters has a large influence on the thermal effects observable in the results. Depending on the heating and cooling rate, effects like glass transition, structural phase transitions, melting and crystallization can be revealed from the thermograms. For the observation of these effects in the most common fiber polymers (polypropylene, polyamide 6 and poly(ethylene terephthalate)), recommendations for experimental parameters are given and the temperature ranges for these effects are indicated.

The development of new spinning processes is demonstrated for poly(vinylidene fluoride), whereas DSC can be used to detect the formation of the piezoelectric β phase. With the help of temperature modulated DSC compared to X-ray diffraction and dynamic mechanical analysis, a process chain for the generation of piezoelectric sensor fibers with possible spinning and drawing parameters was developed. Since the formation of the piezoelectric crystallites can be detected by DSC, it is powerful tool for future development of piezoelectric sensors and actuators.

Electrically conductive nanocomposites based on carbon nanotubes are used as an example for the development of new materials. The presence of nanoparticles influences the crystallization process dependent on the chemical structure of the polymer matrix, whereas two different mechanisms can be detected. The resulting structure then determines the percolation threshold for electrical conductivity. If polymer crystallites can directly grow on the CNT, the particles are separated and percolation threshold is only determined by the particle geometry and orientation. In the other case, the affinity of the polymer chains to the CNT is lower, so that they act as nucleation seeds. CNT can then aggregate and form conductive paths. Due to the higher separation of the two components, percolation threshold is lower.

Author details

W. Steinmann*, S. Walter, M. Beckers, G. Seide and T. Gries
Institut für Textiltechnik (ITA) der RWTH Aachen University,
Aachen, Germany

Acknowledgement

Special thanks to the Deutsche Forschungsgemeinschaft (German research foundation, DFG) for funding the project GR 1311/10-2.

19. References

[1] Engelhardt A (2011) Global synthetic industrial filament yarn and fiber markets. Chemical Fibers International. j. 61: 122.
[2] Fourné F (1995) Synthetische Fasern: Herstellung, Maschinen und Apparate, Eigenschaften; Handbuch für Anlagenplanung, Maschinenkonstruktion und Betrieb. München: Hanser. 880 p.
[3] Gries T, Sattler H (2005) Chemiefasern. In: Winnacker K, Küchler L, editors. Chemische Technik. Weinheim: Wiley
[4] Wulfhorst B, Gries T, Veit D (2006) Textile Technology. München: Hanser. 320 p.

---

* Corresponding Author

[5] Walter S, Steinmann W, Gries T, Seide G, Schenuit H, Roth G (2010) Production of textile fabrics from PVDF multifilament yarns with textile titer. Technical Textiles. j. 53: E95-E97.

[6] Walter S, Steinmann W, Gries T, Roth G, Seide G, Schenuit H. Tools for savers of life : production of warp-knitted fabrics from PVDF multifilament yarns of textile fineness. Kettenwirk-Praxis. j. 45: 34-36.

[7] Walter S, Steinmann W, Seide G, Gries T, Roth G (2012) Development of innovative fibre materials for technical applications : fine polyvinylidene fluoride filaments and fabrics. Filtration. j. 12: 60-64.

[8] Michaeli W (2010) Einführung in die Kunststoffverarbeitung. München: Hanser. 256 p.

[9] Ziabicki A (1976) Fundamentals of fiber formation. London: Wiley. 504 p.

[10] Ziabicki A, Kawai H (1985) High-Speed Fiber Spinning – Science and Engineering Aspects. London: Wiley. 586 p.

[11] Salem D (2000) Structure formation in polymeric fibers. München: Hanser. 578 p.

[12] Beyreuther R, Brünig H (2007) Dynamics of fibre formation and processing. Berlin: Springer. 365 p.

[13] Nakajima T (1994) Advanced fiber spinning technology. Cambridge: Woodhead. 276 p.

[14] Prevorsek D, Oswald H (1990) Melt-spinning of PET and nylon fibers. In: Schultz J, Fakirov F, editors. Solid state behavior of linear polyesters and polyamides. Englewood Cliffs: Prentice Hall.

[15] Chu B, Hsiao B (2001) Small-Angle X-ray Scattering of Polymers. Chemical Reiews. j. 101:1727-1761.

[16] Mandelkern L (2002) Crystallization of Polymers – Volume 1 equilibrium concepts. Cambridge: University Press. 448 p.

[17] Fließbach T (2006) Statistische Physik. München: Elsevier. 408 p.

[18] Strobl G, Cho T (2007) Growth kinetics of polymer crystals in bulk. European Physics Journal. j. 23:55-65.

[19] Banik N, Boyle F, Sluckin T, Taylor P (1979) Theory of structural phase transitions in poly(vinylidene fluoride). Physical Review Letters. j. 43: 456-460.

[20] Strobl G (2007) A multiphase model describing polymer crystallization and melting. Lecture Notes in Physics. j. 714: 481-502.

[21] Frick A, Stern C (2006) DSC-Prüfung in der Anwendung. München: Hanser. 164 p.

[22] Schawe J, Hütter T, Heitz C, Alig I, Lellinger D (2006) Stochastic temperature modulation: A new technique in temperature-modulated DSC. Thermochimica Acta. j. 446:147-155.

[23] Cheng S (2002) Handbook of Thermal Analysis and Calorimetry, Volume 3: Applications to Polymers and Plastics. Amsterdam: Elsevier. 828 p.

[24] Hatakeyama T, Quinn F (2000) Thermal Analysis: Fundamentals and Applications to Polymer Science. London: Wiley.

[25] LyondellBasell (2009) Product Date and Technical Information Moplen HP561R. Rotterdam: LyondellBasell.

[26] BASF Corporation (2011) Datasheet Ultramid B24N03. Ludwigshafen: BASF Corporation.

[27] Invista Resins & Fibers GmbH (2009) Product Specification Polyester Chips 4048. Gersthofen: Invista Resins & Fibers GmbH.

[28] Hsiao B (2011) Polymorphism, Preferred Orientation and Morphology of Propylene-Based Random Copolymer Subjected to External Force Fields (dissertation). State University of New York at Stony Brook.

[29] Liu Y, Cui L, Guan F, Gao Y, Hedin NE, Zhu L, Fong H (2007) Crystalline Morphology and Polymorphic Phase Transitions in Electrospun Nylon 6 Nanofibers. Macromolecules. j. 40:6283-6290.

[30] Drobny J (2001) Technology of fluoropolymers. Boca Raton: CRC Press. 227 p.

[31] Nalwa H (1995) Ferroelectric Polymers – Chemistry, Physics and Applications. New York: Marcel Dekker. 912 p.

[32] Herbert J (1982) Ferroelectric transdurcers and sensors. New York: Gordon and Breach. 464 p.

[33] Lovinger A (1983) Poly(vinylidene fluoride). Developments in Crystalline Polymers. j. 1:195.

[34] Takahashi Y, Matsubara Y, Tadokoro H (1983) Crystal structure of form II of poly(vinylidene fluoride). Macromolecules. j. 16:1588-1792.

[35] Hasegawa R, Takahashi Y, Chatani Y, Tadokoro H (1972) Crystal structures of three crystalline forms of poly(vinylidene fluoride). Polymer Journal. j. 3:600-610.

[36] Weinhold S, Litt M, Lando J (1980) The crystal structure of the gamma phase of poly(vinylidene fluoride). Macromolecules. j. 13: 1178-1183.

[37] Takahashi Y, Tadokoro H (1980) Crystal structure of form III of poly(vinylidene fluoride)

[38] Wang T, Herberg J, Glass A (1988) The Application of Ferroelectric polymers. Glasgow: Blackie and Son. 304 p.

[39] Du C, Zhu B, Xu Y (2007) Effect of stretching on crystalline phase structure and morphology of hard elastic PVDF fibers. Journal of Applied Polymer Science. j. 104:2254-2259.

[40] Steinmann W, Walter S, Seide G, Gries T, Roth G, Schubnell M (2011) Structure, properties, and phase transitions of melt-spun poly(vinylidene fluoride) fibers. Journal of Applied Polymer Science. j. 120:21-35.

[41] Lund A, Hagström B (2010) Melt-spinning of Poly(vinylidene fluoride) Fibers and Influence of Spinning Parameters on β-phase crystallinity. Journal of Applied Polymer Science. j. 116:2685-2693.

[42] Walter S, Steinmann W, Schütte J, Seide G, Gries T, Roth G, Wierach P, Sinapius M (2011) Characterisation of piezoelectric PVDF monofilaments. Materials Technology. j. 26:140-145.

[43] Solvay Solexis (2003) SOLEF 1006 PVDF Homopolymer (data sheet). Brussels: Solvay.

[44] Hougham G (1999) Fluoropolymers – Volume 2 Properties. New York: Kluwer. 408 p.

[45] Grady B (2011) Carbon Nanotube Polymer Composites. Hoboken: Wiley. 352 p.

[46] Lee S, Kim M, Kim S, Youn J (2008) Rheological and electrical properties of polypropylene/MWCNT composites prepared with MWCNT masterbatch chips. European Polymer Journal. j. 44:1620-1630.

[47] Wescott J, Kung P, Maiti A (2006) Conductivity of carbon nanotube polymer composites. Applied Physics Letters. j. 90.

[48] Alig I, Pötschke P, Pegel S, Dudkin S, Lellinger D (2008) Plastic composites containing carbon nanotubes: Optimisation of processing conditions and properties. Rubber Fibre Plastics. j. 3:92-95.

[49] Steinmann W, Walter S, Seide G, Gries T (2011) Melt spinning of electrically conductive bicomponent fibers. In: Adolphe D, Schacher L, editors: 11th World Textile Conference AUTEX 2011, 8-10 June 2011, Mulhouse, France. Book of Proceedings, Volume 2. Mulhouse : Ecole Nationale Supérieure d'Ingenieurs Sud-Alsace. pp. 716-721.

[50] Wulfhorst J, Steinmann W, Walter S, Seide G, Heidelmann M, Weirich T, Gries T (2011) Nanoadditivation of meltspun filament yarns. In: ICONTEX 2011 International Congress of Innovative Textiles, 20-22 October 2011. Çorlu/Tekirdağ : Namik Kemal University. pp. 39-39.

[51] Wulfhorst J, Steinmann W, Walter S, Seide G, Gries T (2011) Antibacterial behaviour and electrical conductivity of textiles by melt spinning of yarns with incorporated nanoparticles. In: Lahlou M, Koncar V, editors: Book of Abstracts / 3rd Edition of the International Conference on Intelligent Textiles and Mass Customisation ITMC'2011, October 27, 28 & 29, 2011, Casablance & Marrakesh, Morocco. Casablanca: ESITH . p. 33.

[52] Skrifvars M, Soroudi A (2009) Melt Spinning of Carbon Nanotube Modified Polypropylene for Electrically Conducting Nanocomposite Fibers. Solid State Phenomena. j. 151:43-47.

[53] Liu K, Sun Y, Lin X, Zhou R, Wang J, Fan S, Jiang K (2010) Scratch-resistant, highly conductive and high-strength carbon nanotubes-based composite yarn. ACS Nano. j. 10:5827-5834.

[54] Kyrylyuk A, Hermant M, Schilling T, Klumperman B, Koning C, van der Schoot P (2011) rolling electrical percolation in multicomponent carbon nanotube dispersions. Nature Nanotechnology. j. 6:364-369.

[55] LyondellBasell (2011) Product Date and Technical Information Moplen HP561R. Rotterdam: LyondellBasell.

Chapter 3

Liquid-Solid Phase Equilibria of Paraffinic Systems by DSC Measurements

Luis Alberto Alcazar-Vara and Eduardo Buenrostro-Gonzalez

Additional information is available at the end of the chapter

1. Introduction

Several industrial sectors around the world deal with paraffinic wax in their processes or make use of it in their products. Hence, understanding physical properties of paraffins is of industrial importance. Some of these industrial sectors are: petroleum production, petroleum refining and products, chemical, energy and consumer products [1]. However, as it has been widely reported in literature [2-6], one of the most affected industrial sectors by the paraffin crystallization phenomena is the petroleum industry. Crude oils contain heavy paraffins that may form solid wax phases at low temperature in the pipelines and hydrocarbon production facilities. The problems caused by wax precipitation decreasing production rates and failure of facilities, are a major concern in the production and transportation of hydrocarbon fluids [7]. Paraffin waxes are mixtures of a wide range of high molecular weight alkanes that can crystallize from crude oils or solutions primarily due to temperature decreasing. They are rather non-polar molecules and their interactions are expected to be van der Waals or London dispersion type [4]. Paraffin waxes consist of branched (iso), cyclic and straight chain (normal) alkanes having chain lengths in excess of 17 carbon atoms (C_{17}) and potentially up to and over C_{100} [8]. However, despite the fact that crude oils are extremely complex systems containing a multitude of components, it is generally accepted that the crystallizing materials that form the deposits are primarily n-alkanes [9-10]. Therefore, in order to obtain a greater insight on the formation of wax deposits to prevent and solve these problems, it is necessary to get a deep knowledge of the mechanisms involved on the n-paraffins crystallization process.

Other industrial problems associated to the paraffin phase behavior have been reported in literature and summarized below. In diesel fuels production operations, fuel-filter plugging and other associated fuel handling problems can occur in cold weather due to paraffin crystallization. Moreover, fuels produced from Fischer–Tropsch syntheses that are currently being investigated for converting natural gas to liquids (fuels) can be particularly problematic due to amounts of higher molecular weight paraffin wax produced [1]. Phase equilibrium data of n-alkane systems with different solvents are of importance for the safe and efficient operation of chemical plants. They are necessary for high-pressure polymerization processes and for the design of oil-recovery processes. Besides its importance for technological processes such as crystallization and purification at high pressure, phase equilibrium properties provides a good tool for examining the thermodynamic nature of many systems [11].Recently, the use of phase change material (PCM) thermal energy storage has gained considerable attention because of its high storage density (amount of energy stored per unit mass), and a narrow temperature range for charging and discharging the storage. Paraffin waxes have been used as PCM for many applications because of their advantageous thermal performances and phase behavior [12]. Finally, the control of crystallization processes is a problem of quite general relevance, which appears in many practical fields such as pharmaceutical and specialty chemical industries [13-15]. "Crystal design or engineering" enables, in principle, a direct handling of the structure, size, and shape of crystals entering into the elaboration of materials. Classical means of controlling size, morphology, and polymorphic expression of crystals make use of parameters such as temperature, pH, supersaturation, and solvent quality [15].

Some experimental techniques reported in literature such as Microscopy and X-ray diffraction are powerful methods to determine the crystal structures but give limited insight into the crystallization process [16], while others methods used to get the liquid-solid equilibrium of paraffins have been used [17-18], but they are very complex due to they require the establishment of the equilibrium at each temperature of interest and the measurement of the composition of the phases present. Finally, visual methods have been also reported to measure solubility and phase behavior of paraffin waxes [1, 19]; however, these methods cannot be applied to test dark samples (e.g. black crude oils) [20]. Hence, for the study and measurement of paraffin crystallization process, Differential scanning calorimetry (DSC) is an experimental method widely used due to its simplicity, accuracy and fast response to monitor the phase transitions during cooling and heating that gives related thermodynamic quantities such as heat capacity and enthalpies of transition [3, 14, 16, 20-23]. DSC has been usually used for the determination of wax appearance and/or dissolution temperatures (WAT or WDT) in petroleum products [3, 20]. The WAT or cloud point is the singularly most important parameter relating to wax formation [4] and it is the temperature at which waxes first crystallize from solution during a cooling process. So that accurate WAT measurements by using reliable methods such as DSC are desirable since it represents a key factor to characterize the wax precipitation phenomena.

The objective of this chapter is to present the use of DSC technique on the measurement and characterization of the liquid-solid phase equilibria of paraffins. First, the details of the DSC method and the experimental conditions used to get the key properties to characterize the liquid-solid phase equilibria of paraffins are described. Then, experimental studies about the effect of the chemical nature of solvent and asphaltenes on liquid-solid phase behavior of paraffinic model systems; are presented and discussed in these sections in order to show specific applications on the use of the Calorimetry to carry out relevant studies of phase equilibria properties. Finally, this chapter presents the characterization of the wax precipitation phenomena by using DSC measurements in crude oils that present solids deposition problems during their production and transporting, where the results obtained by using DSC technique are compared with those obtained with other techniques such as rheometry, spectroscopy and densitometry; in order to show advantages and disadvantages of the use of DSC method to measure liquid-solid phase equilibria of wax in crude oils.

2. DSC methodology applied to measure liquid-solid phase equilibria of paraffins

As it was mentioned above, there are many experimental works in literature [3, 14, 16, 20-23, 28-30] reporting the use of DSC to study the paraffins crystallization process. In this section, it is described the DSC methodology to characterize the liquid-solid phase equilibria of paraffins in model systems and crude oil samples. The objective of this section is to provide the details of the DSC technique and the experimental conditions used to get the key properties that characterize the liquid-solid phase equilibria of paraffins.

The measurement principle of differential scanning calorimetry (DSC) is based on the measurement of the difference in the heat flows to the sample crucible and reference crucible. These heat flows are directly proportional to the temperature difference between the furnace and crucible, but inversely proportional to the thermal resistance of the system. In Figure 1 is shown the measuring cell, furnace and liquid nitrogen cooling chamber of the Shimadzu DSC-60A differential scanning calorimeter used in the experiments to be presented in this work.

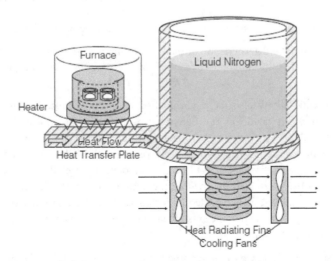

Figure 1. DSC measuring cell and temperature control system (Source: Shimadzu).

The method to obtain the liquid-solid phase equilibrium properties from DSC experiments is explained below. A calibration procedure of the DSC equipment should be performed before carrying out the experiments by using Indium or series of high purity normal paraffins as standard [22]. Each sample (between 10 and 20 mg) is first heated until reaching a temperature higher than expected crystallization onset temperature (WAT) but without reaching the boiling point of the sample. Then, the sample is held isothermally for 1 min., and then cooled to the desired temperature at a pre-defined rate. The cooling/heating rate can be variable; in general low heating/cooling rates would be desirable from an equilibrium point of view [20]. However, by using low cooling rates higher WAT are obtained with a loss of sensitivity to identify the DSC peak onsets, whereas high cooling rates depress measured WAT due to supercooling effects [20]. Differences about ± 1°C on DSC WAT measurements have been observed when using 1, 5 and 10 °C/min as cooling rates in single paraffin solutions [16]; whereas for crude oil mixtures and by using low cooling rates of 0.1 to 1 °C/min, differences about ± 1-2°C were reported [31] Therefore, in the experiments presented in the following sections, we employ a heating/cooling rate of 5 °C/min because it provided sufficient experimental speed and sensitivity to identify onsets of the exo and endothermic peaks. In order to delete any thermal history effects, two heating/cooling cycles are employed, so that crystallization and melting properties are obtained from the second cycle. The crystallization onset temperature (WAT) is determined as the onset of the exothermal peak during the cooling process corresponding to the liquid–solid transition. Under heating conditions, the melting temperature is recorded as the onset of the endothermal peak, whereas the wax disappearance temperature (WDT), temperature at which the last precipitated paraffin re-dissolves in the oil or solution, can be recorded as the endset of the solid–liquid endotherm. Finally, due to that the total energy released during cooling or heating process is proportional to the area between the base line and the exothermal peak or endothermal peak, respectively, the enthalpies of crystallization and melting of the waxy model systems are calculated from the integration of heat flow curve.

In Figure 2 is shown the determination of the equilibrium temperatures (WAT and melting temperature) as well as the enthalpies from the DSC thermograms according to the method explained above.

The DSC technique allows also the determination of the wax precipitation or solubility curve (amount of precipitated wax at different temperatures) as it has been reported [18, 28, 32-33]. It is carried out by assuming that the amount or fraction of precipitated wax in the total wax content is proportional to the percent of accumulated heat released in the total heat released (Crystallization enthalpy), thus the amount of precipitated wax at different temperatures can be determined by dividing the accumulated heat released by the heat of crystallization. This procedure is depicted in Figure 3, where the accumulated heat released for the exothermic peak related to the crystallization of the system 6 wt % of C_{36} in n-decane is plotted as an example [28].

Finally, by using DSC data, we can determine the degree of crystallinity for pure solutes in solvent systems or mixtures by using the following equation [28, 34]:

$$Percent\ crystallinity = [\Delta H_m / \Delta H_{m^\circ}]\ x\ 100 \qquad (1)$$

where ΔH_m is the melting enthalpy of the mixture measured by DSC and ΔH_{m° is the melting enthalpy of the 100% crystalline solute.

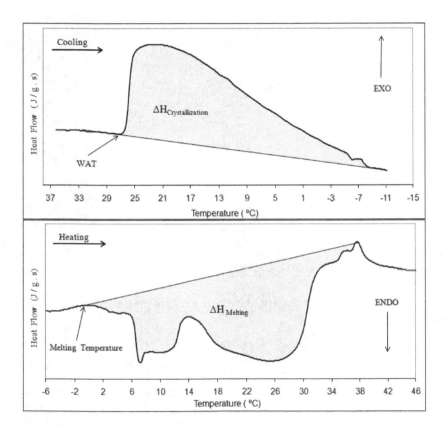

Figure 2. Example of DSC measurements on a paraffinic model system.

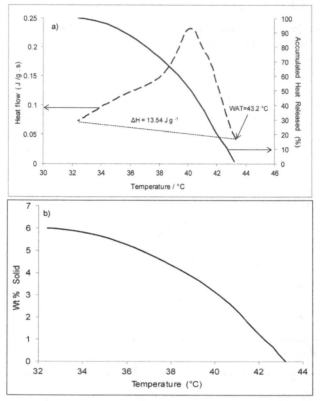

Figure 3. Determination of wax solubility curve of the system 6 wt.% of C₃₆ in *n*-decane: a) accumulated heat released for the DSC exothermic peak and b) wax solubility curve obtained [28].

3. Experimental study of the influence of solvent on paraffin crystallization

The paraffin crystallization process can be influenced by many factors such as paraffin composition, solvent nature, polidispersity, rate of cooling, pressure, kinetics and presence of impurities [1,9, 14, 28, 30, 35-36], so that a better knowledge of factors affecting wax solubility will also improve the understanding of the wax precipitation phenomena in the petroleum industry described above. The studies about the effect of solvent on solubility of waxes reported in literature [1] have shown that waxes do not exhibit ideal solution behavior when crystallizing and that their solubility in a solvent increases as both the solvent molecular size and solvent solubility parameter decrease. The influence of the shape and size of the solvent on solute–solvent interaction and on the n-alkanes solubility has been also described in literature, hence it has been reported that globular or spherical solvents destroy the conformational order in liquid long-chain hydrocarbons [28, 37]. Aromatic solvents have been reported as a help in both inhibiting wax crystal formation and decreasing the amount of the wax deposited [38]. The experimental studies reported in literature have been carried out evaluating the effect of the solvent on cloud point or wax dissolution temperatures. Nevertheless, the wax gelation and deposition processes are actually originated due to the amount of paraffin crystals formed during cooling below WAT. This makes important to evaluate the influence of solvent on the amount of crystallized paraffin at temperatures below WAT. Therefore, in this section is presented an application of the Differential Scanning Calorimetry (DSC) to study the liquid-solid phase behavior of a high molecular weight n-paraffin: hexatriacontane ($C_{36}H_{72}$), in presence of solvents of different chemical nature in order to get a better understanding of the interactions solute-solvent on the paraffin crystallization mechanism.

Figure 4 shows the DSC thermograms of the crystallization and melting behavior of 6 wt. % of hexatriacontane (C_{36}) in different pure and mixed solvents systems. As can be seen, a single and well defined peak is observed during cooling and heating processes, related to the crystallization and melting of the monodisperse sample of the heavy paraffin C_{36} in different solvent systems. However, the endothermic peaks seem to be broader than the exothermic, so while the identification of the crystallization onset temperatures was straightforward; the melting onset temperatures were difficult to identify.

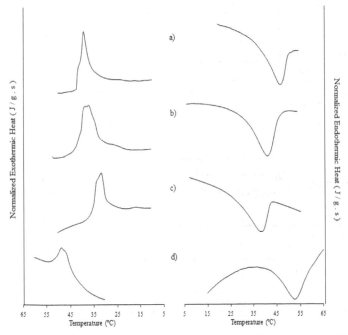

Figure 4. DSC exothermic and endothermic peaks of 6 wt% of C_{36} in different simple and mixed solvents systems: a) 94% of n-decane, b) 47% n-decane + 47% 1-phenyldodecane, c) 47% n-decane + 47% xylene and d) 94 % of squalane [28].

Crystallization and melting properties of the model systems investigated are shown in Table 1. The influence of the solvent aromaticity over the solution of C_{36} in n-decane was studied by adding mono-aromatic solvents: xylene and 1-phenyldodecane. The data show clearly the effect of the solvent chemistry on those properties. Lower values of crystallization and melting enthalpies are obtained in presence of aromatic solvents. The magnitude of the enthalpies decrease is related to the aromaticity of the solvent mixture, where greater aromaticity causes a greater diminishing of crystallization and melting enthalpies. Hence, the aromatic single rings interspersed among hexatriacontane molecules hinder their interactions, preventing an efficient ordering during cooling, then the paraffin crystal networks of a solid phase formed in such circumstance result significantly less ordered, as indicated by the lower values of the crystallinity index calculated from DSC data (see Table 2) for the model systems with aromatic solvents.

Solvent System	WAT (°C)	Enthalpy of Crystallization (J/g)	Melting Temperature (°C)	Enthalpy of Melting (J/g)	WDT (°C)	Solvent system aromaticity [a]
94 % n-decane	43.2	13.54	30.52	12.48	45.6	0
47% n-decane + 47% xylene	37.5	8.05	11.96	9.31	37.92	0.43
47% n-decane + 47% 1-phenyldodecane	47.5	10.66	20.25	10.09	41.82	0.12
94 % squalane	54.5	6.36	37.1	12.56	52.89	0

[a] Calculated as the aromaticity factor of aromatic solvent multiplied by its molar fraction in the mixture, where aromaticity factor of the xylene and 1-phenyldodecane are 0.75 and 0.333 respectively.

Table 1. Crystallization and melting properties of the system 6 % of C_{36} in different solvents systems [28]

Solvent System	Onset Crystallization Temperature (°C)	Endset Crystallization Temperature (°C)	ΔT [a] (°C)	Δt [b] (min)	DSC Crystallinity [c] (%)
94 % n-decane	43.2	32.4	10.08	2.16	7.21
47% n-decane + 47% xylene	37.5	23.5	14	2.8	5.38
47% n-decane + 47% 1-phenyldodecane	47.5	22.5	25	5	5.83
94 % squalane	54.5	41.08	13.42	2.68	7.26

[a] ΔT = Onset - Endset
[b] For a cooling rate = 5 °C/min
[c] For a melting enthalpy of hexatriacontane = 172.9 J/g [39]

Table 2. DSC crystallization data of the system 6 % of C_{36} in different solvent systems [28]

A way to explain the depressing effect of aromatic solvents over the enthalpies of crystallization and melting is through the entropy change that suffers the system during liquid - solid phase transition. It can be assumed that magnitude of the entropy change in the crystallization and melting processes for a paraffinic – aromatic mixture is lower than that for a 100% aliphatic system, like the C_{36} - n-decane system, because of the poor crystal network ordering of the solid phase formed in presence of aromatic solvents, which is reflected in the lower values of their crystallinity index. On another hand, a crystal network with greater disorder introduced by the aromatic molecules of the solvent is weaker, so its melting temperature tends to be lower. The WAT is also lowered due to the presence of aromatic solvents; however, in the case of the 1-phenyldodecane, a less aromatic solvent than xylene and with a greater molecular weight than decane, its aliphatic chain of 12 carbons has a significant "ordering" effect in the paraffinic crystal network that surpasses the effect of its aromatic rings regards to the depressing of the WAT and melting temperature observed with the xylene.

In contrast, by changing the solvent system from n-decane to squalane (a C_{24} aliphatic chain with six methyl branches), the WAT is significantly increased to 54.5 °C with a dramatic depression of the enthalpy of crystallization. This behavior is influenced by the size and structure of the solvent, in a similar way to that observed for the 1-phenyldodecane – n-decane solvent system. The methyl branches of the C_{24} iso-paraffin inhibits the efficient ordering of C_{36} molecules during crystallization process, forming a less ordered solid phase than that formed in C_{36}-decane system, which is reflected in the low enthalpy of crystallization measured (6.36 J/g), on the other hand, the highest WAT observed is a consequence of the greater size of squalane with respect to decane. In fact, as has been reported, solubility decreases (greater WAT) as the solvent size increases due to the inability of the bigger solvent molecule to effectively contact and solvate the solute [1]. Furthermore, despite the disorder in the crystalline arrangement caused by the six methyl branches of the squalane, the effect of its size (24 carbon length) results in the highest values of both temperature and enthalpy of melting.

Table 1 shows also the WDT of C_{36} in different solvents during melting process. Under ideal conditions, the values of WAT and WDT should be similar; however, as has been reported [22] differences between both values can be attributed to the experimental uncertainty and kinetic effects (e.g. supercooling). Our DSC results showed differences in the range between 0.88-5.38 °C which could be attributed to the heating/cooling rate used of 5 °C/min. The effect of solvent chemistry on WDT of C_{36} was similar to that observed for the melting temperature discussed above.

The results presented before showed a significant influence of solvent chemistry on crystallization and melting of C_{36}; however, in order to get a better understanding of the paraffin crystallization process in presence of solvents of different chemical structure at temperatures below WAT, solubility curves were obtained by using the DSC data as can be seen in Figure 5. As expected, the crystallization process of C_{36} in squalane starts before respect to the other systems, as a consequence of solvent chain length, followed by the other mixtures according to their respective WATs.

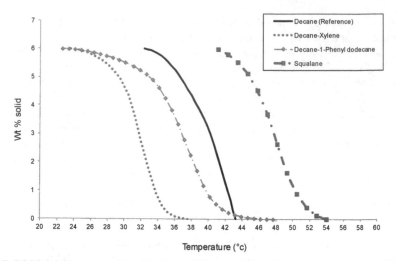

Figure 5. DSC Solubility curves of 6 wt % of C₃₆ in different simple and mixed solvents systems [28].

The chemical structure of the solvent affects significantly the crystallization rate of the Hexatriacontane C₃₆. Table 2 shows that the crystallization process is slower for the systems with aromatic solvents. An evident consequence of this is the fact that lower amounts of solids are formed in these systems with respect to the decane system below its WAT as can be observed in Figure 5; this can be attributed to the aromatic ring interfering with the normal crystal growth, retarding the conformational ordering of C₃₆ molecules in the solid phase created.

The degree of crystallinity obtained by DSC for the solid phase formed in presence of squalane, shown in Table 2, points out the effect of competition between the size and branching of the aliphatic solvents in the crystallinity of the solid phases formed. Due to the greater size of squalane with regard to the *n*-decane, it would be expected a greater crystallinity index for the solid phase formed in presence of squalane; however its ramifications limit the possibility of achieving an efficient conformational ordering of the crystal network in the solid phase, which results in a crystallinity index value similar to that obtained for the *n*-decane system.

The results presented in this section showed that the solvent aromaticity was a key factor that results in inhibition of the paraffin crystallization process, decreasing WAT, by promoting the creation of a solid phase partially disordered due to the presence of aromatic single rings interspersed among paraffin molecules hindering their efficient ordering during the cooling process.

4. DSC study of the effect of asphaltenes on liquid-solid phase equilibria

Asphaltenes are the most heavy and polar fraction in the crude oil. The asphaltene fraction is formed by many series of relatively large molecules containing aromatic rings, several heteroaromatic and napthenic ring plus relatively short paraffinic branches [8]. Some authors have proposed that asphaltenes form aggregates with a core formed by aromatic regions, and aliphatic chains on the periphery interacting with the surrounding oil [40-41]. The aliphatic portions of the asphaltene permit an interaction of asphaltenes with waxes. A handful of studies have evaluated the asphaltene-wax interactions and their effect on wax crystallization and gelation properties [28-30, 42-45]. The study of the influence of the asphaltenes and their chemical nature on the wax crystallization phenomena of paraffinic model systems; is presented in this section. DSC measurements allowed getting crystallization properties of paraffinic model systems in presence of asphaltenes of different origin with different molecular structure.

The effect of asphaltenes was studied on liquid-solid equilibrium of the binary system: tetracosane ($C_{24}H_{50}$) - octacosane ($C_{28}H_{58}$). The composition used in these model systems was 15 and 10 wt% of C_{24} and C_{28} respectively in the following solvent systems: a) 75% of decane, b) mixture of 37.5 % of decane + 37.5 % of xylene and c) mixture of 37.5 % of decane + 37.0 % of xylene + 0.5 % of asphaltenes. The asphaltene samples used in this study, labeled as AsphPC and AsphIri, were extracted from two Mexican crude oils of the southern region. These asphaltenes of different chemical nature were characterized by using elemental analysis, vapor pressure osmometry, ^1H and ^{13}C NMR spectroscopy in order to get their molecular parameters. Details of the experimental techniques used for the characterization of these asphaltenes and their effect on phase-equilibrium and rheological properties of waxy model systems have been reported recently in literature [28, 30]. Some of the molecular parameters of these asphaltenes are shown in Table 3. As can be seen, the AsphPC asphaltenes are more aromatic than AsphIri asphaltenes, and its aromatic core is also bigger and more condensed, whereas the aromatic core of AsphIri asphaltenes is richer in alkyl substituents comprising methyl groups, alkyl chains and naphthenic rings [30]. DSC exothermic peaks of the model systems are plotted in Figure 6 and their crystallization properties are shown in Table 4.

Symbol	Definition	Asphaltene sample	
		AsphIri	AsphPC
R_A	Aromatic rings	7.09	20.69
f_a	Aromaticity factor	0.50	0.67
ϕ	Condensation index	0.53	0.71
n	Average number of carbon atoms per alkyl substituents	5.99	5.47
σ	Aromatic substitution index	0.55	0.37
n_{ac}	Average length of the alkyl chains	11.84	11.12

Table 3. Molecular parameters of AsphPC and AsphIri asphaltenes [30]

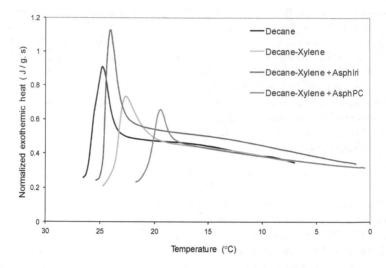

Figure 6. DSC exothermic peaks of the binary system C_{24}-C_{28} in different solvent systems and with asphaltenes of different chemical nature [28].

Solvent System	WAT (°C)	Enthalpy of Crystallization (J/g)	Endset (°C)	ΔT[a] (°C)	Δt[b] (min)
Decane	26.5	37.09	7	19.5	3.9
Decane-Xylene	24.81	36.51	6.09	18.72	3.74
Decane-Xylene + AsphIri	25.66	58.42	1	24.66	4.93
Decane-Xylene + AsphPC	21.77	33.01	0.53	21.24	4.24

[a] ΔT = WAT - Endset; [b] For a cooling rate = 5 °C / min

Table 4. DSC crystallization data of the binary system C_{24}-C_{28} in different solvent systems and with asphaltenes [28]

As can be observed, the presence of an aromatic solvent as the xylene decreases slightly both WAT and crystallization enthalpy due to the disorder-effects generated by the aromaticity as was discussed before. However, crystallization properties were notably affected by the presence of asphaltenes where their chemical nature played an important role. It has been reported that flocculated asphaltenes providing nucleation sites for waxes increase WAT [44], but also it has been reported [30] that asphaltenes decreased very slightly the WAT. According with other studies [45] the effect of the asphaltenes on the WAT depends of the aggregation state of the asphaltenes.

In these model systems, the results obtained showed a slight increasing of WAT due to presence of the aliphatic asphaltenes (AsphIri) and a decreasing in presence of the more aromatic asphaltenes (AsphPC) with respect to the C_{24}-C_{28} in decane-xylene model system without asphaltenes. For another hand, the crystallization heat increases significantly in presence of AsphIri asphaltenes, whereas diminishes moderately in presence of AsphPC asphaltenes. These results make evident the effect of the chemical nature of different asphaltenes over the crystallization behavior of paraffinic systems. A greater abundance of aliphatic chains in AsphIri asphaltenes permits a better interaction with the paraffins of the C_{24}-C_{28} system promoting co-crystallization phenomena, where the asphaltenes are partially integrated to the crystal network and probably acting as nucleation sites, causing a slight increasing of WAT. Moreover, the partial immobilization of the paraffins engaged in the interactions with the asphaltene alkyl chains may promote a *"quasicrystallization"* phenomenon of the paraffins in the asphaltene network, such interaction results in exothermic effects as has been reported in literature [46, 47], which explains the significant increasing of crystallization heat of the model system with AsphIri asphaltenes as observed in Table 4. On the other hand, the most aromatic asphaltenes (AsphPC) with a bigger and more condensed aromatic core and with a smaller amount of aliphatic substituents inhibit in some extent the paraffin-asphaltene interactions so that they cannot be incorporated to the paraffin crystal structure hindering nucleation process and crystal network growth, which results in a WAT decreasing with a lower crystallization enthalpy due to the formation of a disordered solid phase.

The wax solubility curves of these systems are plotted in Figure 7, as can be seen the effect of solvent and asphaltenes of different chemical nature is evident. Regarding to the C_{24} - C_{28} mixture in *n*-decane, the presence of xylene reduces moderately both the amount of solid formed and the WAT, although the temperature interval of their solubility curves are very similar. However, the effect of asphaltenes on wax solubility curve is very significant considering its low concentration in the model system (0.5%), particularly in the case of the more aromatic AsphPC asphaltene. As in the case of C_{36} in different solvents, the rate of wax precipitation for the C_{24}-C_{28} mixture in xylene is significantly affected by the presence of a small amount of highly aromatic compounds such as the asphaltenes. The data in Table 4 shows also that the presence of asphaltenes in the paraffinic system increases around 25% the time required to precipitate the total of paraffins (Δt) for a cooling rate of 5°C/ min.

These results showed that asphaltenes are practically acting in this system as inhibitors of the paraffins precipitation. When the asphaltenes have a structure highly condensed with a certain degree of aromatic substitution, that allows some kind of interaction with the paraffins, the inhibition effect is greater due the steric interference and the disorder generated in the paraffinic network which difficult the molecular recognition among paraffin molecules avoiding the growth of stable crystalline networks and therefore the formation of the solid phase. Otherwise a less condensed aromatic structure with a greater substitution degree have a better interaction with paraffins and thus, it could play a role as nucleation site increasing both WAT and the amount of solid phase formed, at least in a certain temperature range as can be observed in Figure 7, but even in such case the disruptive effect that introduces the aromatic core of the asphaltenes prevails inhibiting the crystallization process, as it was observed for the model system with the AsphIri asphaltenes. These phenomena are sketched in Figure 8.

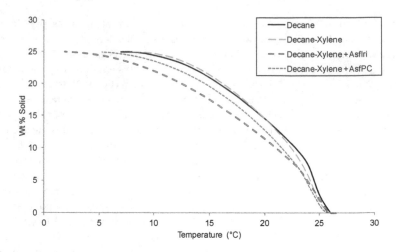

Figure 7. Wax solubility curves of the binary system C₂₄-C₂₈ in different solvent systems and with asphaltenes of different chemical nature [28].

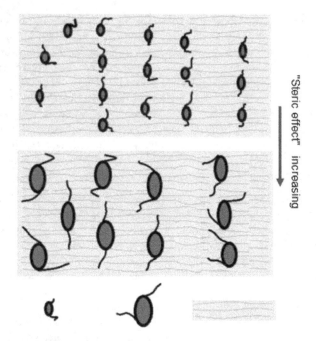

Asphaltene Asphlri   Asphaltene AsphPC   Paraffin network

Figure 8. Schematic representation of the "Steric effect" of asphaltenes of different chemical nature on paraffin crystallization.

5. Wax precipitation study in crude oils by DSC measurements

With the ongoing trend in deep water developments, flow assurance has become a major technical and economic issue. The avoidance or remediation of wax deposition is one key aspect of flow assurance [3-4, 6-7]. In order to develop solutions to the wax deposition problem is necessary to get a deep understanding of the crystallization phenomena in which the crude oil composition, particularly the content of high molecular weight paraffins and asphaltenes have a significant impact [29]. Comparison of experimental methods for measurement of wax precipitation in crude oils have been reported in literature [5, 21, 24-27, 29], where wax detection limits vary depending on the measurement technique, oil composition, thermal history, time of measurement and fluid properties related to crystal nucleation and growth [27]. In this section is presented a characterization of the wax precipitation phenomena in crude oils by using DSC measurements. Wax appearance temperature is measured by using DSC and these results are compared with those obtained by using other techniques such as rheometry, spectroscopy and densitometry in order to show advantages and disadvantages of the use of DSC method to measure liquid-solid phase equilibria of waxes in crude oils. The importance to get the wax melting temperature and crystallinity degree by DSC is analyzed also as key parameters to evaluate the propensity of crude oils to present wax precipitation problems during crude oil production and transporting. Three Mexican crude oils of the southern region labeled as RDO1, RDO2 and J32 are studied in this work. Crude oils RDO1 and RDO2 present wax precipitation and deposition problems during their production and transportation, whereas crude oil J32 presents a severe asphaltene precipitation and deposition problem along the well during primary production. The cloud point temperatures of the crude oils and the crystallization and melting properties of their isolated waxes were determined by using the DSC method according to the procedures and conditions described and presented in section II of this chapter. However, in these experiments for waxes, under heating conditions, the wax melting point was recorded as their endothermic peak temperature.

In Figure 9 are shown the DSC thermograms of crude oils during cooling process. As it has been reported [22], during cooling, a decrease of the solvating power of the oil matrix results in precipitation of solid particles of wax.

The thermograms of crude oils RDO1 and RDO2 show one well defined exothermal peak from which the WAT can be easily determined 19.2 and 18.5°C respectively. In contrast the thermogram of crude oil J32 presents two exothermal peaks, first one well defined, a liquid-solid phase transition whose onset corresponding to a WAT of 28°C and a second one broad not well defined, with an onset around 20.8°C.

Other methods were employed to determine the WAT of the crude oil samples, these experimental techniques were fourier transform infrared spectroscopy (FT-IR), rheometry and densitometry. Figure 10 shows the WAT determination by these methods described briefly below. FT-IR method is based on the fact that the absorbance between the wave numbers 735 and 715 cm^{-1} attributed to the rocking vibrations of the long chain methylene (LCM) groups (the major component of the solid wax formed in crude oils), has been found to increase with a decrease in temperature due to the formation of a solid phase [48]. Below the WAT, the higher absorptive of the solid phase made up of LCM grups, contributes strongly to the total absorbance, which gives rise a change in the slope of the plot of the area of the absorbance peak (735 to 715 cm^{-1}) of each FT-IR spectrum as a function of the temperature. The temperature where the change in slope occurs is recorded as the WAT as observed in Figure 10 a). Rheometric WAT measurements were carried on the basis that petroleum fluids exhibit non-Newtonian behavior below WAT and Newtonian behavior above the WAT which follows the Arrhenius temperature dependence:

$$\mu = A \, e^{\, Ea / RT} \tag{2}$$

where μ is the Newtonian dynamic viscosity, A is the Arrhenius pre-exponential factor, Ea is the activation energy of viscous flow, R is the universal gas constant and T is the absolute temperature. The formation and growing of solid wax crystals dispersed in the crude oil medium causes a viscosity increasing during a cooling process [35]. In this way, from viscosity-temperature curves, WAT is recorded as the temperature of the deviation of the Arrhenius law as it is shown in Figure 10 b). Finally, the WAT determination by using densitometry method is carried out by identifying the temperature at which a sharp change in the slope of density – temperature curve obtained during a cooling process that become evident the onset of wax crystallization as can be observed in Figure 10 c).

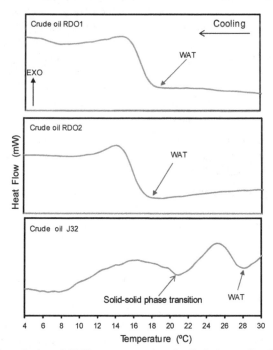

Figure 9. Exothermic peaks from DSC Thermograms of crude oils during cooling [29].

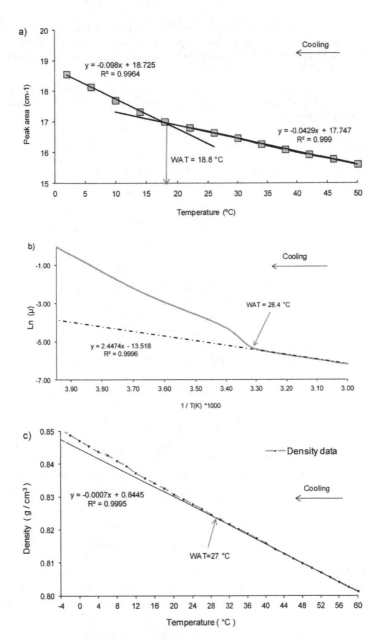

Figure 10. WAT determination by using different experimental methods: a) FT-IR WAT determination for crude oil RDO1, b) Rheometric WAT measurement for crude oil RDO2 from Arrhenius plot and c) Density-Temperature profile for crude oil RDO1 [29].

Table 5 shows the WAT measurements by using the different techniques for the crude oils
investigated in this work. According to these results, good agreement is obtained between
DSC and FT-IR method in the WAT measurement, whereas Rheometry method apparently
overestimated cloud point for crude oils RDO1 and RDO2 and underestimate it in the case
of crude oil J32. WAT detection difficulties related with effects of crude oil composition
were found when using Densitometry method where the uncertainty associated with the
WAT measurement was high.

Crude oil	WAT (°C)				Average (°C)	Deviation (%)
	DSC	Rheometry	Densitometry	FT-IR		
RDO1	19.2	32.1	27	18.8	24.2	± 26.5
RDO2	18.5	28.4	21	16.5	21.1	± 24.6
J32	28	16.3	*ND	30.7	25	± 30.6

*Not detectable

Table 5. WAT measurements of crude oils [29]

The good agreement between DSC and FT-IR methods can be explained in terms of their
high sensitivity to the energetic variations related to the phase transition phenomena of
paraffins. The Rheometry and Densitometry methods, however, need that a critical amount
of solid wax come out of solution to produce a detectable change in the rheological
properties or density of the crude oil sample and thus to identify the WAT. On the other
hand, the WAT values of the crude oils RDO, measured by the rheometric technique were
higher than DSC values ; hence liquid-liquid demixing effects not detected by DSC and FT-
IR methods during cooling can have a significant impact on rheological behavior of crude
oil samples resulting in an overestimation of WAT by the rheometric technique.

Waxes isolated from crude oils were also analyzed by DSC experiments in order to get their
crystallization and melting properties. Figure 11 shows the exothermic and endothermic
peaks obtained from DSC thermograms of the crude oil waxes. As can be observed, during
cooling and heating, a single and well-resolved peak was obtained for each of the RDO1 and
RDO2 waxes, whereas exo and endothermic peaks of Wax J32 were broader and partially
well defined between 60 and 6 °C.

The crystallization and melting properties of waxes characterized by DSC are shown in
Table 6. It can be seen that wax J32 has the highest crystallization temperature (60.2 °C)
followed by wax RDO2 (52.3 °C) and wax RDO1 (50.1 °C). In the case of crystallization
enthalpies, the lowest was obtained for wax J32 and the highest for wax RDO1. As expected
and due to the crystallizing materials that form the waxy deposits are primarily *n*-alkanes,
there is a correlation between wax crystallization temperature and crude oil WAT (see
Figure 12).

The wax melting temperature is an important parameter that can be used to define the
temperature at which pipe walls or storage facilities may need to be heated in order to
remove solid deposits [49]. The lowest melting temperature found in wax J32 (35.9 °C) helps

to minimize the propensity of crude oil J32 to present wax precipitation problems during production and transportation, despite of its highest both crystallization temperature whereas crude oils RDO1 y RDO2, whose wax fractions have the highest melting temperatures present wax precipitation and deposition problems in the well head and downstream. A correlation between temperatures and enthalpies of melting was observed in waxes analyzed by DSC, wax RDO1 with highest melting temperature (45.8 °C) has also the highest melting enthalpy (112.89 J/g). This correlation found in this work is in agreement with previous results reported in literature [50]. Finally in Figure 13 is plotted the relationship between wax melting temperature and wax crystallinity, where a good correlation can be observed. In fact and as expected, the lowest degree of crystallinity of wax J32 indicates that its crystal structure is weaker and less stable, with a higher disorder degree, and thus making it easier to melt as pointed out its lower melting temperature.

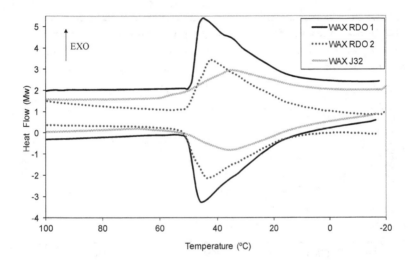

Figure 11. Exo and endothermic peaks from DSC thermograms of waxes [29].

Property	Wax		
	RDO1	RDO2	J32
Crystallization Temperature (°C)	50.1	52.3	60.2
Crystallization Enthalpy (J/g)	110.8	107.11	82.29
Melting Temperature (°C)	45.8	43.6	35.9
Enthalpy of melting (J/g)	112.89	110.56	102.13
DSC Crystallinity (%)	70.29	68.84	62.46

Table 6. Crystallization and melting properties of waxes [29]

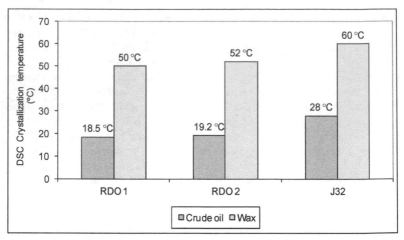

Figure 12. Corrrelation between DSC crystallization temperatures of crude oils and their wax fraction

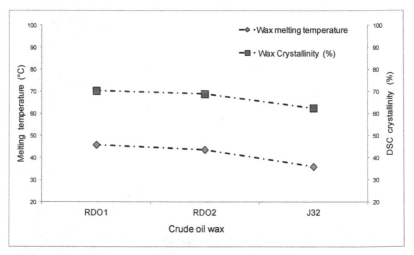

Figure 13. Corrrelation between wax melting temperature and crystallinity.

As conclusions of this section is established that from the methods presented here for the WAT determination, DSC technique is recommended for the identification of phase transitions in crude oils due to its high sensitivity and reliability. Crude oils RDO1 and RDO2, whose wax fractions have the highest melting temperatures and crystallinity degree present wax precipitation and deposition problems in the well head and downstream, whereas the crude oil J32 that has the wax fraction with the lowest melting temperature and crystallinity degree do not present wax deposition problems during production. DSC measurements allowed identifying key parameters useful to evaluate the wax precipitation in crude oil samples.

6. Conclusions

In this chapter we have showed the usefulness of the DSC technique on the characterization of the liquid-solid phase equilibrium of paraffins. It was shown that DSC is an experimental method widely used due to its simplicity, accuracy and fast response to monitor phase transitions. Good capabilities and advantages of the DSC method were shown in order to carry out relevant studies on thermodynamic properties of paraffins. In this chapter it was shown that DSC measurements allowed elucidating the interactions of paraffins with solvents and asphaltenes of different chemical nature during the crystallization and melting phenomena. DSC technique was very useful on the characterization of the wax precipitation phenomena in crude oils due to its high sensitivity and reliability for the identification of phase transitions and crystallinity measurement, which allows evaluating tendency of crude oils to present wax precipitation problems in production and transporting operations. Finally, in order to develop solutions to the wax formation problems presented in some industrial sectors such as petroleum industry is necessary to get a deep understanding of the paraffin crystallization phenomena where the DSC method can give us valuable information as was shown in this chapter.

Author details

Luis Alberto Alcazar-Vara and Eduardo Buenrostro-Gonzalez
Instituto Mexicano del Petróleo, Programa Académico de Posgrado. Eje Central Lázaro Cárdenas, México, D.F.

Acknowledgement

The authors express gratitude to the Instituto Mexicano del Petróleo (IMP) for both providing facilities and granting permission to publish results. L.A.A.V thanks CONACYT and the Programa Académico de Posgrado of IMP for the economic support granted during his Ph.D. studies. The authors thank to Mr. J.A. Garcia-Martinez from IMP for his excellent assistance on asphaltenes characterization.

7. References

[1] Jennings DW, Weispfennig K (2005) Experimental solubility data of various n-alkane waxes: Effects of alkane chain length, alkane odd versus even carbon number structures and solvent chemistry on solubility. Fluid Phase Equilibria. 227:27–35.
[2] Burger ED, Perkins TK, Striegler JH (1981) Studies of Wax Deposition in the Trans. Alaska Pipeline. J. Petroleum Tech. 33:1075-86.
[3] Rønningsen HP, Bjørndal B, Hansen AB, Pedersen WB (1991) Wax precipitation from North Sea oils. 1. Crystallization and dissolution temperature, and Newtonian and non-Newtonian flow properties. Energy & Fuels. 5:895–908.

[4] Leontaritis KJ (1995) The Asphaltene and Wax Deposition Envelopes. The Symposium on Thermodynamics of Heavy Oils and Asphaltenes, Area 16C of Fuels and Petrochemical Division, AIChE Spring National Meeting and Petroleum Exposition, Houston, Texas, March 19-23.

[5] Hammami A, Raines MA (1999) Paraffin Deposition from Crude Oils: Comparison of Laboratory Results to Field data. SPE Journal. 4(1):9-18.

[6] Alboudwarej H, Huo Z, Kempton E (2006) Flow-Assurance Aspects of Subsea Systems Design for Production of Waxy Crude Oils. SPE Annual Technical Conference and Exhibition in San Antonio, Texas, USA. 24-27 September.

[7] Coutinho JAP, Edmonds B, Moorwood T, Szczepanski R, Zhang X (2006) Reliable wax predictions for flow assurance. Energy & Fuels. 20:1081–1088.

[8] Speight JG (1999) The Chemistry and Technology of Petroleum (3rd ed.), Marcel-Dekker, New York.

[9] Garcia MD, Orea M, Carbognani L, Urbina A (2001) The Effect of Paraffinic Fractions on Crude Oil Wax Crystallization. Petroleum Science and Technology. 19(12):189–196.

[10] Garcia MD, Urbina A (2003) Effect of Crude Oil Composition and Blending on Flowing Properties. Petroleum Science and Technology. 21(5-6):863–878.

[11] Domanska U, Morawski P (2005) Influence of Size and Shape Effects on the High-Pressure Solubility of n-Alkanes: Experimental Data, Correlation and Prediction. Journal of Chemical Thermodynamics. 37:1276-1287.

[12] He B, Martin V, Setterwall F (2004) Phase Transition Temperature Ranges and Storage Density of Paraffin Wax Phase Change Materials. Energy. 29: 1785–1804.

[13] Giron D (2002) Applications of Thermal Analysis and Coupled Techniques in Pharmaceutical Industry. Journal of thermal analysis and Calorimetry. 68: 335-357.

[14] Senra M, Panacharoensawad E, Kraiwattanawong K, Singh P, Fogler HS (2008) Role of n-Alkane Polydispersity on the Crystallization of n-Alkanes from Solution. Energy & Fuels. 22:545–555.

[15] Marie E, Chevalier Y, Eydoux F, Germanaud L, Flores P (2005) Control of n-Alkanes Crystallization by Ethylene-Vinyl Acetate Copolymers. Journal of Colloid and Interface Science. 290(2):406-418.

[16] Guo X, Pethica BA, Huang JS, Prud'homme RK (2004) Crystallization of Long-Chain n-Paraffins from Solutions and Melts As Observed by Differential Scanning Calorimetry. Macromolecules. 37:5638–5645.

[17] Van Winkle TL, Affens WA, Beal EJ, Mushrush GW, Hazlett RN, DeGuzman J (1987) Determination of Liquid and Solid Phase Composition in Partially Frozen Middle Distillate Fuels. Fuel. 66(7):890-896.

[18] Coutinho JAP, Ruffier-Meray V (1997) Experimental Measurements and Thermodynamic Modeling of Paraffinic Wax Formation in Undercooled Solutions. Ind. Eng. Chem. Res. 36:4977– 4983.

[19] Bhat NV, Mehrotra AK (2004) Measurement and Prediction of the Phase Behavior of Wax-Solvent Mixtures: Significance of Wax Disappearance Temperature. Ind. Eng. Chem. Res. 43(12): 3451-3461.

[20] Hammami A, Ratulowski J, Coutinho JAP (2003) Cloud Points: Can We Measure or Model them? .Petroleum Science and Technology. 21(3&4):345-358.

[21] Hammami A, Raines MA (1999) Paraffin Deposition from Crude Oils: Comparison of Laboratory Results to Field data. SPE J. 4(1): 9-18

[22] Hansen AB, Larsen E, Pedersen WB, Nielsen AB (1991) Wax precipitation from North Sea crude oils. 3. Precipitation and dissolution of wax studied by differential scanning calorimetry. Energy & Fuels. 5:914–923.

[23] Guo X, Pethica BA, Huang JS, Adamson DH, Prud'homme RK (2006) Effect of Cooling Rate on Crystallization of Model Waxy Oils with Microcrystalline Poly(ethylene butane). Energy & Fuels. 20:250–256.

[24] Monger-McClure TG, Tackett JE, Merrill LS (1999) Comparisons of Cloud Point Measurement and Paraffin Prediction Methods .SPE Production & Facilities. 14(1):4–10.

[25] Lira-Galeana C, Hammami A (2000) Wax Precipitation from Petroleum Fluids: A Review. in: Asphaltenes and Asphalts 2. Yen TF, Chilingarian G eds. Elsevier Science Publishers: Holland. pp. 557-608.

[26] Leontaritis KJ, Leontaritis JD (2003) Cloud Point and Wax Deposition Measurement Techniques. SPE Paper No. 80267, SPE International Symposium on Oilfield Chemistry, Houston, Texas. February 5-8.

[27] Coutinho JAP, Daridon JL (2005) The Limitations of the Cloud Point Measurement Techniques and the Influence of the Oil Composition on its Detection. Petroleum Science and Technology. 23: 1113-1128.

[28] Alcazar-Vara LA, Buenrostro-Gonzalez E (2012) Experimental Study of the Influence of Solvent and Asphaltenes on Liquid-Solid Phase Behavior of Paraffinic Model Systems by using DSC and FT-IR Techniques. Journal of Thermal Analysis and Calorimetry. 107(3): 1321-1329.

[29] Alcazar-Vara LA, Buenrostro-Gonzalez E (2011) Characterization of the Wax Precipitation in Mexican Crude Oils. Fuel Processing Technology. 92(1): 2366–2374.

[30] Alcazar-Vara LA, Garcia-Martinez JA, Buenrostro-Gonzalez E (2012) Effect of Asphaltenes on Equilibrium and Rheological Properties of Waxy Model Systems. Fuel. 93: 200–212.

[31] Svetlichnyy DS, Didukh AG, Aldyarov AT, Kim DA, Nawrocki M, Baktygali AA (2011) Study of Heat Treatment and Cooling Rate of Oil Mixtures Transported by "Kumkol-Karakoin-Barsengir-Atasu" Pipeline. Online journal Oil and Gas Business. Available: http://www.ogbus.ru/eng/authors/Svetlichnyy. Accessed 2012 April 13.

[32] Han S, Huang Z, Senra M, Hoffmann R, Fogler HS (2010) Method to Determine the Wax Solubility Curve in Crude Oil from Centrifugation and High Temperature Gas Chromatography Measurements. Energy & Fuels. 24:1753-1761.

[33] Coto B, Martos C, Peña JL, Espada JJ, Robustillo MD (2008) A New Method for the Determination of Wax Precipitation from non-Diluted Crude Oils by Fractional Precipitation. Fuel. 87: 2090-2094.

[34] Conti DS, Yoshida MI, Pezzin SH, Coelho LAF (2006) Miscibility and Crystallinity of Poly(3-hydroxybutyrate)/Poly(3-hydroxybutyrate-co-3-hydroxyvalerate) Blends. Thermochimica Acta. 450:61–66.

[35] Paso K, Senra M, Yi Y, Sastry AM, Fogler HS (2005) Paraffin Polydispersity Facilitates Mechanical Gelation. Ind. Eng. Chem. Res. 44:7242-7254.

[36] Vieira LC, Buchuid MB, Lucas EF (2010) Effect of Pressure on the Crystallization of Crude Oil Waxes. II. Evaluation of Crude Oils and Condensate. Energy & Fuels. 24:2213-2220.

[37] Domanska U, Morawski P (2005) Influence of Size and Shape Effects on the High-Pressure Solubility of n-Alkanes: Experimental Data, Correlation and Prediction. Journal of Chemical Thermodynamics. 37:1276-1287.

[38] Rakotosaona R, Bouroukba M, Petitjean D, Dirand M (2008) Solubility of a Petroleum Wax with an Aromatic Hydrocarbon in a Solvent. Energy & Fuels. 22:784–789.

[39] Dirand M, Bouroukba M, Chevallier V, Petitjean D, Behar E, Ruffier-Meray V (2002) Normal Alkanes, Multialkane Synthetic Model Mixtures, and Real Petroleum Waxes: Crystallographic Structures. Thermodynamic Properties and Crystallization. J. Chem. Eng. Data. 47:115–143.

[40] Mullins OC, Betancourt SS, Cribbs ME, Dubost FX, Creek JL, Andrews AB, Venkataramanan L (2007) The Colloidal Structure of Crude Oil and the Structure of Oil Reservoirs. Energy & Fuels. 21(5):2785-2794.

[41] Carbognani L, Rogel E (2003) Solid Petroleum Asphaltenes Seem Surrounded by Alkyl layers. Petroleum Science and Technology. 21(3-4):537-556.

[42] Garcia MD, Carbognani L (2001) Asphaltene-Paraffin Structural Interactions. Effect on Crude Oil Stability. Energy & Fuels. 15(5):1021-1027.

[43] Venkatesan R, Ostlund JA, Chawla H, Wattana P, Nyden M, Fogler HS (2003) The Effect of Asphaltenes on the Gelation of Waxy Oils. Energy & Fuels. 17(6):1630-1640.

[44] Kriz P, Andersen SI (2005) Effect of Asphaltenes on Crude Oil Wax Crystallization. Energy & Fuels. 19(3):948-953.

[45] Tinsley JF, Jahnke JP, Dettman HD, Prud'home RK (2009) Waxy Gels with Asphaltenes 1: Characterization of Precipitation, Gelation, Yield Stress, and Morphology. Energy & Fuels. 23(4):2056-2064.

[46] Mahmoud R, Gierycz P, Solimando R, Rogalski M (2005) Calorimetric Probing of n-Alkane-Petroleum Asphaltene Interactions. Energy & Fuels. 19:2474 –2479.

[47] Stachowiak C, Viguie J-R, Grolier J-P, Rogalski M (2005) Effect of n-Alkanes on Asphaltene Structuring in Petroleum Oils. Langmuir. 21:4824-4829.

[48] Roehner RM, Hanson FV (2001) Determination of Wax Precipitation Temperature and Amount of Precipitated Solid Wax versus Temperature for Crude Oils using FT-IR Spectroscopy. Energy & Fuels. 15:756–763.

[49] Flow assurance design guideline (2001) Deepstar IV project.

[50] Alghanduri LM, Elgarni MM, Daridon JL, Coutinho JAP (2010) Characterization of Libyan Waxy Crude Oils. Energy & Fuels. 24:3101–3107.

Indirect Calorimetry to Measure Energy Expenditure

Chapter 4

Energy Expenditure Measured by Indirect Calorimetry in Obesity

Eliane Lopes Rosado, Vanessa Chaia Kaippert and Roberta Santiago de Brito

Additional information is available at the end of the chapter

1. Introduction

Obesity is a multifactorial disease characterized by excessive deposition of fat in adipose tissue, which may be due to excessive energy intake, and or changes in body energy expenditure, resulting in positive energy balance [1].

Mika Horie et al [2] demonstrated that obese women had higher total energy expenditure (TEE), compared with normal weight. However, this increase may be due to increased basal metabolic rate (BMR) due to higher fat-free mass (FFM) and energy demand during physical activity. However, Melo et al [3] found that obese individuals are economical, the metabolic point of view. Therefore, the energy expenditure (EE) per kilogram of body weight at the give time is lower in obese individuals.

The low metabolic rate, expressed relative to FFM seems to be a risk factor for weight gain [4]. In a prospective study in Pima Indians, Ravussin et al. [5] showed that both the low resting metabolic rate (RMR) and low TEE increased risk of weight gain. The basal energy expenditure (BEE) and resting (REE) can be obtained through BMR and RMR, respectively, multiplied by 24 hours (1440 minutes).

There are several methods for the assessment of EE with different levels of precision, including indirect calorimetry, which measures the metabolic rate by the determination of oxygen consumption (O_2) (with a spirometer), the production of carbon dioxide (CO_2) and excretion of urinary nitrogen, for a given period of time [6]. This technique relies on the fact that all the O_2 consumed and CO_2 produced is due to the oxidation of the three major energy substrates, which are fats, carbohydrates and proteins [7].

Recognizing the need to estimate EE in institutions that have no indirect calorimetry, researchers have proposed the use of specific equations, developed from calorimetry studies in groups of individuals with similar clinical characteristics [8]. Although the estimate of EE is the most common method, the predictive equations might generate errors [9]. Shetty [10] considers that the equations used to estimate the BEE in normal weight adults have reasonable precision (coefficient of variation 8%).

In clinical practice it is impracticable to measure the calorimetric methods for EE, so the international use of the equations was recommended, modified from a compilation of data carried out by Schofield [11]. Studies conducted in different ethnic groups found that these equations provide high BEE estimates, particularly for residents in the tropics [12-14]. Wahrlich and Anjos [14] confer these differences to the fact that equations have been developed mostly from population samples of North America and Europe which show differences in body composition, and live in different environmental conditions.

It is known that in populations with severe obesity is actually more difficult to fit the equations, because there is the difficulty in choosing the weight to be applied in the equation, which may influence a lot the results [15]. The use of current weight leads to the overestimation of the results independent of the equation to be applied, and the use of ideal or adjusted weight can result in the underestimation of energy needs [16].

Considering that obesity is a chronic disease of epidemiological importance, nutritional intervention studies have been developed in order to propose strategies for the prevention and treatment of this disease. The chronic imbalance between energy intake and EE results in positive energy balance and body weight gain. One way of evaluating the influence of dietary components in the EE, it through the measurement of energy metabolism by indirect calorimetry.

Obesity is also considered multifactorial, with genetic and environmental causes. In this sense, also studied the influence of candidate genes to obesity in metabolic variables, and indirect calorimetry is used in these studies.

The purpose of this chapter is to assess the importance of indirect calorimetry in the assessment of EE in obese individuals, both in study of nutrition intervention and influence of genes in EE, and in the validation and adequacy of existing prediction equations, which were not created for this population.

2. Indirect calorimetry

Indirect calorimetry remains a gold standard in measuring EE in the clinical settings. Indirect calorimetry offers a scientifically-based approach to customize a patient's energy needs and nutrient delivery to maximize the benefits of nutrition therapy. With recent advances in technology, indirect calorimeters are easier to operate, more portable, and affordable. Increased utilization of indirect calorimetry would facilitate individualized patient care and should lead to improved treatment outcomes [17].

Indirect calorimetry is considered a standard method, after validation by comparison with the direct calorimetry [18], however, its use is restricted to research due to the demanding cost and time for its conclusion [14], requiring the use of prediction equations in clinical practice.

According to Green (1994) [19], this technique is based on the principles that there are no considerable reserves of O_2 in the body, the O_2 uptake reflects the oxidation of nutrients, that all the chemical energy in the body comes from the oxidation of carbohydrates, fats and proteins, and that the ratio of O_2 consumption and CO_2 produced for the oxidation of these macronutrients are fixed. After determining the concentration of O_2 inspired and CO_2 expired, the calculation of RMR can be done with the equation of Weir (1949) [20, 21].

Given the difficulties associated with urine collection 24 hours, and found little difference between the results using the complete formula and simplified (2%), many authors have chosen to use the equation of Weir (1949) disregarding urinary losses of nitrogen [22].

The amount of O_2 used for oxidation and CO_2 production will depend on the substrate being oxidized. The respiratory quotient (RQ = Volume CO_2/Volume O_2) varies with the nutrients are being oxidized [14]. The table below describes the values of RQ complexes corresponding to the use of energy substrates [22].

RQ	Interpretation
0.67	Ethanol oxidation
0.71	Fat oxidation
0.82	Protein oxidation
0.85	Oxidation of a mixed diet
1.0	Carboydrate oxidation
1.0 – 1.2	Lipogenesis

Adapted of Materese (1997) [22].

Table 1. Values for interpretation of respiratory quotient (RQ) according to substrate oxidation.

The RQ is divided into non-protein RQ, which reflects the participation of carbohydrates and fats, and protein RQ, which represents the use of proteins. The rate of protein oxidation is obtained by determining the amount of nitrogen excreted in the urine during the test [7, 8].

The evaluation of RMR should only be initiated after a period of rest to minimize possible effects of recent physical activities such as dressing, driving or walking. In practice, it is recommended around twenty minutes of rest, for longer periods can cause the individual to sleep or get impatient [14].

During the measurement of RMR, the ambient temperature should be maintained in the neutral thermal zone, ie between 25 and 26°C, by Henry & Emery (1986) [14]. In order to evaluate the effects of temperature on EE, Dauncey (1981) [23] subjected women to direct calorimetry for thirty (30) hours at 22°C and 28°C. The evaluation of the RMR morning for 30 minutes, after twelve hours of fasting, demonstrated that the production of heat at the lowest temperature (22°C) was significantly greater (mean of 11 ± 3.2%) compared to the higher temperature (28ºC). Wahrlich & Anjos (2001) [14] point out that at ambient temperatures below the neutral zone, the use of protective clothing may be sufficient to prevent the increase in EE caused by the cold.

Compher et al. (2006) [24] conducted a systematic literature review to determine the optimal conditions for obtaining reliable measures of RMR by indirect calorimetry. Based on this survey, the following information was highlighted by the authors: 1) food, ethanol, caffeine and nicotine affect RMR for a variable number of hours after consumption, so its use should be controlled prior to measurement, 2) daily activities increased the RMR, however, a short rest period (20 minutes) before the test is sufficient, 3) moderate or vigorous physical activities presented most impact on the metabolism and therefore should be avoided in hours prior to measurement of RMR; 4) the measure with a duration of ten minutes with the first five minutes disregarded, and the remaining five minutes with a coefficient of variation of up to 10% guarantee accurate measurements of RMR, 5) the trial site should be physically comfortable and should be evaluated for ten to twenty minutes of rest before the start of measurement.

For research on assessment of RMR recommended: fasting for at least 6 hours, abstaining from caffeine during the night, nicotine and alcohol for at least 2 hours, of moderate physical activity for at least 2 hours, and vigorous physical activity for 14 hours [24].

In relation to the proper position for holding the indirect calorimetry, the authors emphasize that the most important is to ensure that each individual is physically comfortable during the test, and that measures are always taken in the same position. However, they warn that some positions require increased muscle tone and, therefore, could influence the measurement of RMR, and for research that has attempted to evaluate the RMR should keep the individual in the supine posture with or slightly higher [24].

The time required for obtaining an accurate measurement of RMR is only about five to ten minutes, discarding the first five minutes, provided that no changes occur in over 10% VO_2 and VCO_2 and 5% in RQ. Accordingly, one measure would be sufficient to describe the RMR for twenty-four hours. However, if you cannot guarantee the stability of the readings, the combination of two or three repetitions increase the precision of the measurements for the extrapolation to twenty-four hours [8, 24].

For studies involving analysis of the thermic effect of food (TEF), the indirect calorimetry should be conducted for a period of 6 hours, as measured by shorter periods are not able to fully assess the TEF [24].

With regard to environmental characteristics, it is recommended that the room temperature is comfortable, that is, between 20 and 25°C, the environment is quiet, with soft lighting and humidity control [24].

In females, should be avoided that the energy metabolism assessments are performed during the luteal phase because this phase of the menstrual cycle are described changes such as water retention, increased of body weight and energy demand, changes in lipid profile and metabolism of some nutrients (vitamin D, calcium, magnesium and iron), emotional hypersensitivity, aches and changes in eating behavior [25].

So, indirect calorimetry is a simple and affordable tool for measuring EE and for quantifying the utilization of macronutrients. Its use is becoming increasingly widespread, but it is necessary to know its methodological features and its theoretical and practical limitations. Indirect calorimetry measures the rate of REE, the major component of the TEE. Coupling the measurement of body composition to that of REE expands the diagnostic potential of indirect calorimetry. Once the lean and fat compartments have been measured, it is possible to establish on the basis of REE whether an individual is hyper- or hypometabolic. The clinical applications are practically unlimited [26].

Although the basic principles of indirect calorimetry are well established, it is important to recognize that there are several potential pitfalls in the methodology and data interpretation that must be appreciated to properly understand and apply the results derived from this technique. One must recognize that the fundamental measurement provided by indirect calorimetry is the net disappearance rate of a substrate regardless of the metabolic interconversions that the substrate may undergo before its disappearance from its metabolic pool. Under most circumstances, direct oxidation represents the major route by which a substrate disappears from its metabolic pool, and the two terms are often used interchangeably. However, under conditions when rates of gluconeogenesis, ketogenesis, or lipogenesis are elevated, the presumed equivalence between oxidation and disappearance may no longer apply, even though the actual measurements derived from indirect calorimetry remain valid. When indirect calorimetry is combined with other in vivo metabolic techniques (e.g., the insulin clamp or radioisotope turnover methods) it can provide a powerful tool for noninvasively examining complex metabolic processes [27].

Indirect calorimetry can also evaluate BEE. The main difference between REE and BEE is that REE is measured after the individual dislocation to the exam site, necessitating the prior resting period of 30 minutes to neutralize the effects of the physical activity performed [28]. Study found that REE is 10-15% higher than the BEE [29].

3. Utilization of indirect calorimetry in obesity

3.1. Evaluation of prediction equations

Kross et al [30] evaluated the accuracy of multiple regression equations to estimate REE in critically ill patients, especially for obese patients. A total of 927 patients were identified, including 401 obese patients. There were bias and poor agreement between measured REE and REE predicted by the Harris-Benedict, Owen, American College of Chest Physicians, and Mifflin equations ($p > 0.05$). There was poor agreement between measured and predicted REE by the Ireton-Jones equation, stratifying by sex. Ireton-Jones was the only equation that was unbiased for men and those in weight categories 1 and 2. In all cases except Ireton-Jones, predictive equations underestimated measured REE. The authors concluded that none of these equations accurately estimated measured REE in this group of mechanically ventilated patients, most underestimating energy needs. The authors concluded that is necessary to develop predictive equations for adequate assessment of energy needs.

Ullah et al [31] compared measured REE using the bedside with indirect calorimetry commonly used prediction equations, considering that the accuracy of prediction equations for estimating REE in morbidly obese patients is unclear. A total of 31 morbidly obese patients (46 kg/m²) were studied. Pre-operative REE with indirect calorimetry was measured and compared with estimated REE using the Harris-Benedict and Schofield equations. All patients subsequently underwent a Roux-en-Y gastric bypass and were repeated measurements at six weeks and three months following surgery. The equations overestimated REE by 10% and 7%, by Harris-Benedict and Schofield equations, respectively. After weight loss the difference between the estimated and measured REE reduced to 1.3%. The accuracy improved after surgery induced weight loss, confirming their validity for the normal weight population. The study demonstrated that indirect calorimetry should be used in morbid obesity.

Cross-sectional study developed by our research group (unpublished data) with 92 women (35.60 ± 6.66 years) with excess body weight (34.41 ± 4.71 kg/m²), Brazilian and Spanish. This study assessed the women in a metabolic unit, after fasting for 12 hours without performing strenuous physical activity in the last 24 hours and with minimal effort. The evaluation was performed using the open-circuit respiratory hood with indirect calorimetry (Deltatrac Metabolic Monitor-R3D) [6]. For the calculation of EE, it was used the values of the following volumes; inspired O_2 (VO₂), expired CO_2 (VCO₂) (ml / min) and urinary nitrogen [6-28], obtained by the calorimeter. In Brazilian women, it was found that the estimates obtained by Harris-Benedict, Shofield, FAO / WHO /ONU and Henry & Rees did not differ from REE of indirect calorimetry, which presented higher values than the equations proposed by Owen, Mifflin-St Jeor and Oxford. In Spanish women, also the equations proposed by Owen, Mifflin-St Jeor and Oxford presented EE lower than the indirect calorimetry, while the other equations did not differ from the indirect calorimetry. Both are women, Brasilian and Spanish, the best equations were FAO / WHO / ONU, Harris-Benedict Shofield and Henry & Rees.

Study aimed to validate the published predictive equations for REE in 76 normal weight (44.8 kg, 19.0 kg/m²) and 52 obese (64.0 kg, 25.9 kg/m²) Korean children and adolescents in the 7-18 years old age group. The open-circuit indirect calorimetry using a ventilated hood system was used to measure REE. Sixteen REE predictive equations were included, which were based on weight and/or height of children and adolescents, or which were commonly used in clinical settings despite its use based on adults. For the obese group, the Molnar, Mifflin, Liu, and Harris-Benedict equations provided the accurate predictions of > 70% (87%, 79% 77%, and 73%, respectively). On the other hand, for non-obese group, only the Molnar equation had a high level of accuracy (bias of 0.6%, RMSPE of 90.4 kcal/d, and accurate prediction of 72%). The accurate prediction of the Schofield (W/WH), WHO (W/WH), and Henry (W/WH) equations was less than 60% for all groups [32].

Alves et al. (2009) [33] compared the RMR obtained by indirect calorimetry with predict equations (Harris-Benedict (HB) and Ireton-Jones (IJ)) in 44 patients with excess body weight. The nearest RMR in fasting was obtained with the equation HB using the current body weight (1.873 $\pm$ 484 kcal / day and 1798 $\pm$ 495 kcal / day for HB and indirect calorimetry, respectively). However, the authors emphasize the need to employ the indirect calorimetry for the determination of EE of obese, because despite the similarity found between the absolute REE measured by indirect calorimetry and the prediction equation, there are significant ranges of variability, suggesting that the ideal method and more accurate to obtain the actual REE in this population is the indirect calorimetry.

3.2. Evaluation of the effect of nutritional interventions and obesity candidate genes in energy expenditure

Study with 60 obese women (34.59 ± 7.56 years) was conducted in order to evaluate the influence of fat diet and peroxisome proliferator-activated (PPARγ2) and β2-adrenergic receptor genes on energy metabolism. It was found that polymorphism in PPARgamma2 resulted in increased in fat oxidation, regardless of genotype of β2-adrenergic receptor gene. Polyunsaturated fatty acids (PUFA) intake can assist in weight loss, but the genotype of the genes assessed determines the type of fat that should be ingested [34].

The same research group developed another study with sixty obese women (30–46 years) which were divided into two groups depending on the genotype of PPARγ2 (Pro12Pro and Pro-12Ala/Ala12Ala). At baseline and after two nutritional (short- or long-term) interventions, measurement of anthropometrical and body composition (bioelectrical impedance) variables, dietary assessments, energy metabolism (indirect calorimetry) measurements as well as biochemical and molecular (PPARγ2 genotype) analyses were performed. All women received a high-fat test meal to determine the post-prandial metabolism (short term) and an energy-restricted diet for 10 weeks to determine the effect of diet in long term. The Pro12Ala polymorphism in the PPARγ2 gene influenced energy metabolism in the assayed short- and long-term situations since the response to both nutritional interventions differed according to the genotype. The results suggest that fat oxidation and EE may be lower in Pro12Pro carriers compared to Pro12Ala/Ala12Ala genotypes, while in obese women with Pro12Ala/Ala12Ala polymorphisms in the PPARγ2 gene fat oxidation was negatively correlated with the monounsaturated fatty acids (MUFA) and PUFA (%) intake [35].

The difference in structure of fatty acids, including the chain length, degree of unsaturation and the position of the double bond, can affect the rate of oxidation of fatty acids. Medium chain saturated fatty acids (MCSFA) are more easily oxidized than the long chain saturated fatty acids (LCSFA), while unsaturated fatty acids (UFA) are more easily oxidized compared to saturated chain acids (SFA) with the same chain length [36].

Using indirect calorimetry studies also indicate that the PUFA shows a higher oxidation compared to SFA, both in men and in obese normal [37, 38]. Piers et al (2002) [39] found that changes in the type of fat dietary may have a beneficial effect on reducing body weight in men who consume high fat content, since the oxidation rate postprandial (assessed by Indirect calorimetry) of nutrient is increased after a high MUFA meal, compared with the SFA.

Casas-Agustench et al (2009) [40] aimed to compare the acute effects of three fatty meals with different fat quality on postprandial thermogenesis and substrate oxidation. Evaluated twenty-nine healthy men aged between 18 and 30 years in randomized crossover trial, comparing the thermogenic effects of three isocaloric meals: high in PUFA from walnuts, high in MUFA from olive oil, and high in SFA from fat-rich dairy products. Indirect calorimetry was used to determine RMR, RQ, 5-H postprandial EE and substrate oxidation. Five hours postprandial thermogenesis was higher by 28% after the high PUFA meal (p = 0.039) and by 23% higher after the high MUFA meal (p = 0.035), compared with the high SFA meal. Increased fat oxidation rates no significantly after the two meals rich in UFA and decreased non significantly after the high SFA meal. Postprandial RQ, carbohydrate and protein oxidation measures were similar among meals. The authors concluded that fat quality determined the thermogenic response to a fatty meal but clear effects on substrate oxidation.

Another study was conducted to evaluate whether postprandial abnormalities of EE and / or lipid oxidation are present in healthy, normal-weight individuals with a strong family history of obesity and thus at high risk to become obese. They conducted a case-control study. A total of 16 healthy young men participated in the study. Eight individuals had both parents overweight (father's and mother's body mass index > 25 kg / m^2) and eight had both parents with normal body weight (father's and mother's body mass index < 25 kg / m^2). The group of individuals with overweight parents was similar to that with normal-weight parents (control group) in terms of body mass index and FFM. EE was measured by indirect calorimetry, and blood samples were taken for the evaluation of metabolic variables in the fasting state and every hour for 8 h after a standard fat-rich meal (protein 15%, 34% carbohydrate, 51 fat %, 4090 kJ). Fasting and postprandial EE, and fasting fat and carbohydrate oxidation were both in similar groups. On the contrary, postprandial carbohydrate oxidation (incremental area under curve) was significantly higher and that of fat oxidation lower in the group of individuals with overweight parents. They concluded that normal-weight individuals with a strong family history of obesity present a reduced fat oxidation in the postprandial period. These metabolic characteristics may be considered the early predictors of weight gain and are genetically determined probably [41].

Differences in meal-induced thermogenesis and macronutrient oxidation between lean (n = 19) and obese (n = 22) women after consumption of two different isocaloric meals, one rich in carbohydrate (CHO) and one rich in fat were examined. Women were studied on two occasions, one week apart. In one visit they consumed a CHO-rich meal and in the other visit a fat-rich meal. The two meals were isocaloric and were given in random order. REE and macronutrient oxidation rates were measured and calculated in the fasting state and every hour for 3 h after meal consumption. Meal-induced thermogenesis was not different between lean and obese subjects after the CHO-rich (p = 0.89) or fat-rich (p = 0.32) meal, but it was significantly higher after the CHO-rich compared with the fat-rich meal in the lean and the obese individuals (p < 0.05). Protein oxidation rate increased slightly but significantly after the test meals in both groups (p < 0.01). Fat oxidation rate decreased after consumption of the CHO-rich meal (p < 0.001), whereas it increased after consumption of the fat-rich meal in both groups ($p < 0.01$). CHO oxidation rate increased in both groups after consumption of the CHO-rich meal ($p < 0.001$). Oxidation rates of protein, fat, and CHO during the experiment were not significantly different between lean and obese participants. In conclusion, it was verified that meal-induced thermogenesis and macronutrient oxidation rates were not significantly different between lean and obese women after consumption of a CHO-rich or a fat-rich meal [42].

In a parallel-arm, long-term feeding trial, 24 lean and 24 overweight participants received a daily peanut oil load in a milk shake equivalent to 30% of their REE for eight weeks to evaluate the effects of peanut oil intake on appetite, EE (indirect calorimetry at baseline and week 8), body composition, and lipid profile. Energy intake increased significantly in the overweight but not in the lean participants. A statistically significant body weight gain (median 2.35 kg) was also observed among the overweight subjects, although this corresponded to only 43% of the theoretical weight gain. In the overweight participants, the REE was significantly increased by 5% over the intervention, but no significant difference was observed in the lean subjects. As expected, REE was significantly higher in the overweight than in the lean participants. No marked differences of appetite were observed over time in either group or between overweight and lean participants. These data indicate that ingestion of peanut oil elicits a weaker compensatory dietary response among overweight compared with lean individuals. Body weight increased, albeit less than theoretically predicted [43].

The effects of a moderate-fat diet, high in MUFAs, and a low-fat (LF) diet on EE and macronutrient oxidation before and after a 6-mo controlled dietary intervention were compared. Twenty-seven overweight (body mass index 28.1 ± 0.4 kg/m^2) nondiabetic subjects (18–36 years) followed an 8-wk low-calorie diet and a 2-wk weight-stabilizing diet and then were randomly assigned to a MUFA ($n = 12$) or LF ($n = 15$) diet for 6 mo. Substrate oxidation and 24-h EE were measured by whole-body indirect calorimetry. The first measurement (0 mo) was taken during the weight-stabilizing diet, and the second measurement was taken after the 6-mo intervention. A tendency was seen toward a lower 24-h EE with the MUFA than with the LF diet ($p = 0.0675$), but this trend did not remain after adjustment for the initial loses of fat mass and FFM ($p = 0.2963$). Meal-induced thermogenesis was significantly ($p < 0.05$) lower with the MUFA than with the LF diet. Despite a slightly lower meal-induced thermogenesis, the MUFA diet had an effect on 24-h EE that was not significantly different from that of the LF diet after a 6-mo controlled dietary intervention [44].

Study with 24 healthy, overweight men (body mass index between 25 and 31 kg/m^2) compared the effects of diets rich in medium-chain triglycerides (MCTs) or long-chain triglycerides (LCTs) on body composition, EE, substrate oxidation, subjective appetite, and *ad libitum* energy intake. At baseline and after four weeks of each dietary intervention, EE was measured using indirect calorimetry. Average EE was 0.04 ± 0.02 kcal/min greater ($p < 0.05$) on day 2 and 0.03 ± 0.02 kcal/min (not significant) on day 28 with functional oil (64.7% MCT oil) compared with olive oil consumption. Similarly, average fat oxidation was greater ($p = 0.052$) with functional oil compared with olive oil intake on day 2 but not day 28.

Consumption of a diet rich in MCTs results in greater loss of adipose tissue compared with LCTs, perhaps due to increased EE and fat oxidation observed with MCT intake [45].

A controlled randomized dietary trial was conducted with 26 overweight or moderately obese men and women (body mass index 28-33 kg/m^2) to test the hypothesis that n-3-polyunsaturated fatty acids (n-3-PUFA) lower body weight and fat mass by reducing appetite and *ad libitum* food intake and/or by increasing EE. Diets were administered in an isocaloric fashion for 2 weeks followed by 12 weeks of *ad libitum* intake. The n-3-PUFA and control diets were identical in all regards except for the fatty acid composition. Both groups lost similar amounts of weight when these diets were consumed *ad libitum* for 12 weeks [mean (SD): -3.5 (3.7) kg in the control group vs. -2.8 (3.7) kg in the n-3-PUFA group, $F(1,24)$ = 13.425, $p = 0.001$ for time effect; $F(1,24) = 0.385$, $p = 0.541$ for time × group interaction]. No differences were founds between the n-3-PUFA and control groups with regard to appetite as measured by visual analogue scale, *ad libitum* food intake or, REE as measured by indirect calorimetry, diurnal plasma leptin concentrations, or fasting ghrelin concentrations. These results suggest that dietary n-3-PUFA do not play an important role in the regulation of food intake, EE, or body weight in humans [46].

4. Conclusion

Indirect calorimetry is useful technique in the metabolic evaluation of obese individuals. Despite some methodological limitations, is still the best way to estimate this variable in this population, which is useful both in studies of dietary intervention, intended to propose new strategies for prevention and treatment of obesity, and for validation of predictive equations for energy expenditure in this population.

Abbreviations

BEE - basal energy expenditure
CHO – carbohydrate
EE - energy expenditure
FFM - fat-free mass
LCSFA - long chain saturated fatty acids
LCT - long-chain triglycerides
LF – low-fat
MCSFA - Medium chain saturated fatty acids
MCT - medium-chain triglycerides
MUFA - monounsaturated fatty acids
PPARγ2 - peroxisome proliferator-activated
PUFA - polyunsaturated fatty acids
REE – resting energy expenditure
RMR - resting metabolic rate
RQ - respiratory quotient
SFA – saturated fatty acids

TEE - energy expenditure
TEF - thermic effect of food
UFA - unsaturated fatty acids
VO_2 - inspired O_2
VCO_2 - expired CO_2

Author details

Eliane Lopes Rosado, Vanessa Chaia Kaippert and Roberta Santiago de Brito
Nutrição e Dietética Departament, Federal University of Rio de Janeiro, Rio de Janeiro, Brasil

Acknowledgement

This work was supported by Conselho Nacional de Pesquisa (CNPq) and Fundação de Amparo à Pesquisa do Estado de Minas Gerais (FAPEMIG).

We thank the Federal University of Viçosa and Navarra University.

5. References

[1] Flatt J-P (2007) Differences in Basal Energy Expenditure and Obesity. Obesity. 15(11):2546-8.

[2] Mika Horie L, González MC, Raslan M, torrinhas R, Rodrigues NL, Verotti CC, Cecconello I, Heymsfield SB, Waitzberg DL (2009) Resting energy expenditure in white and non-white severely obese women. Nutr Hosp. 24(6):676-81.

[3] Melo CM, Tirapegui J, Ribeiro SML (2008) Gasto energético corporal: conceitos, formas de avaliação e sua relação com a obesidade. Arq Bras Endocrinol Metab. 52(3):452-464.

[4] Astrup A, Buemann B, Toubro S, Ranneries C, Raben A (1996) Low resting metabolic rate in subjects predisposed to obesity: a role for thyroid status. Am J Clin Nutr. 63(6):879-83.

[5] Ravussin E, Lillioja S, Knowler WC, Christin L, Freymond D, Abbott WG, Boyce V, Howard BV, Bogardus C (1988) Reduced rate of energy expenditure as a risk factor for body-weight gain. N Engl J Med 1988;318(8):467-72.

[6] Ferrannini E (1988) The theoretical bases of indirect calorimetry: a review. Metabolism. 37(3):287-301.

[7] Jéquier E, Acheson K, Schutz Y (1987) Assessment of energy expenditure and fuel utilization in man. Ann Rev Nutr. 7:187-208.

[8] Diener JRC (1997) Calorimetria indireta. Rev Ass Med Brasil. 43(3):245-53.

[9] Frankenfield D, Roth-Yousey L, Compher C (2005) Comparison of Predictive Equations for Resting Metabolic Rate in Healthy Nonobese and Obese Adults: A Systematic Review. J Am Diet Assoc. 105:775-89.

[10] Shetty P (2005) Energy requirements of adults. Public Health Nutrition. 8(7A):994–1009.

[11] Schofield WN (1985) Predicting basal metabolic rate, new standards and review of previous work. Hum Nutr Clin Nutr. 39 suppl 1:5-41.

[12] Piers LS, Diffey B, Soares MJ, Frandsen SL, McCormack LM, Lutschini MJ, O'Dea K (1997) The validity of predicting the basal metabolic rate of young Australian men and women. Eur J Clin Nutr. 51:333-7.

[13] Cruz CM, Silva AF, Anjos LA (1999) A taxa metabólica basal é superestimada pelas equações preditivas em universitárias do Rio de Janeiro, Brasil. Arch Latinoam Nutr. 49(3):232-7.

[14] Wahrlich V, Anjos LA (2001) Aspectos históricos e metodológicos da medição e estimativa da taxa metabólica basal: uma revisão da literatura. Cad Saúde Pública. 17(4):801-817.

[15] Rodrigues AE, Mancini MC, Dalcanale L, Melo ME, Cercato C, Halpern A (2010) Padronização do gasto metabólico de repouso e proposta de nova equação para uma população feminina brasileira. Arq Bras Endocrinol Metab. 54(5):470-476.

[16] Weg MWV, Watson JM, Klesges RC, Clemens LHE, Slawson DL, MCClanahan BS (2004) Development and cross-validation of a prediction equation for estimating resting energy expenditure in healthy African-American and European-American women. Eur J Clin Nutr. 58:474-80.

[17] Haugen HA, Chan LN, Li F [2007] Indirect calorimetry: a practical guide for clinicians. Nutr Clin Pract. 22(4):377-88.

[18] Benedetti FJ, Bosa VL, Mocelin HT, Paludo J, Mello ED, Fischer GB (2011) Gasto energético em adolescentes asmáticos com excesso de peso: calorimetria indireta e equações de predição. Rev Nutr. 24(1): 31-40.

[19] Green JH (1994) Assessment of energy requirements. In: Heatley RV, Green JH, Losowsky MS, editors. Consensus in Clinical Nutrition. Cambridge: Cambridge University Press. pp. 22-37.

[20] Weir JB (1949) New methods for calculating metabolic rate with special reference to protein metabolism. J Physiol. 109:1-9.

[21] Delany JP, Lovejoy JC (1996) Energy expenditure. Endocrinol Metab Clin North Am. 25(4): 831-846.

[22] Matarese LE (1997) Indirect calorimetry: technical aspects. J Am Diet Assoc. 97:S154-S160.

[23] Dauncey MJ (1981) Influence of mild cold on 24 h energy expenditure, resting metabolism and diet induced thermogenesis. Br J Nutr. 45:257-267.

[24] Compher C, Frankenfield D, Keim N, Roth-Yousey L (2006) Best practice methods to apply to measurement of resting metabolic rate in adults: a systematic review. J Am Diet Assoc. 106(6):881-903.

[25] Sampaio HAC (2002) Aspectos nutricionais relacionados ao ciclo menstrual. Rev Nutr. 15(3):309-317.

[26] Battezzati A, Viganò R [2001] Indirect calorimetry and nutritional problems in clinical practice. Acta Diabetol. 38(1):1-5.

[27] Simonson DC, DeFronzo RA [1990] Indirect calorimetry: methodological and interpretative problems. Am J Physiol. 258(3 Pt 1):E399-412.

[28] Kamimura MA, Avesani CA, Draibe SA, Cuppari L (2008) Gasto energético de repouso em pacientes com doença renal crônica. Rev Nutr. 21(1):75-84.

[29] Poehlman ET, Horton ES (2003) Necessidades Energéticas: Avaliação e Necessidades em Humanos. In: Shils ME, Olson JÁ, Shike M, Ross AC. Tratado de Nutrição Moderna na Saúde e na Doença. São Paulo: Ed Manole, 2003. Vol 1. 9º edição.

[30] Kross EK, Sena M, Schmidt K, Stapleton RD (2012) A comparison of predictive equations of energy expenditure and measured energy expenditure in critically ill patients. J Crit Care. 14; [Epub ahead of print].

[31] Ullah S, Arsalani-Zadeh R, Macfie J [2012] Accuracy of prediction equations for calculating resting energy expenditure in morbidly obese patients. Ann R Coll Surg Engl. 94(2):129-32.

[32] Kim MH, Kim JH, Kim EK (2012) Accuracy of predictive equations for resting energy expenditure (REE) in non-obese and obese Korean children and adolescents. Nutr Res Pract. 6(1):51-60.

[33] Alves VGF, Rocha EEM, Gonzalez MC, Fonseca RBV, Silva MHN, Chiesa CA (2009) Assessment of resting energy expenditure of obese patients: comparison of indirect calorimetry with formulae. Clin Nutr. 28:299-204.

[34] Rosado EL, Bressan J, Hernández JAM, Martins MF, Cecon PR (2006) Efecto de la dieta e de los genes PPARγ2 y β2-adrenérgico en el metabolismo energético y en la composición corporal de mujeres obesas. Nutr Hosp. 21(3):317-331.

[35] Rosado EL, Bressan J, Hernández JAM, Lopes IM (2010) Interactions of the PPARγ2 polymorphism with fat intake affecting energy metabolism and nutritional outcomes in obese women. Ann Nutr Metab. 57(3-4):242-250.

[36] DeLany JP, Windhauser MM, Champagne CM, Bray GA (2000) Differential oxidation of individual dietary fatty acids in humans. Am J Clin Nutr. 72(4):905-911.

[37] Jones PJ, Schoeller DA (1988) Polyunsaturated:saturated ratio of diet fat influences energy substrate utilization in the human. Metabolism. 37:145–151.

[38] Jones PJ, Ridgen JE, Phang PT, Birmingham CL (1992) Influence of dietary fat polyunsaturated to saturated ratio on energy substrate utilization in obesity. Metabolism. 41:396–401.

[39] Piers LS, Walker KZ, Stoney RM, Soares MJ, O'Dea K (2002) The influence of the type of dietary fat on postprandial fat oxidation rates: monounsaturated (olive oil) vs saturated fat (cream). Int J Obes. 26:814–821.

[40] Casas-Agustench P, López-Uriarte P, Bullo M, Ros E, Gómez-Flores A, Salas-Salvadó J (2009) Acute effects of three high-fat meals with different fat saturations on energy expenditure, substrate oxidation and satiety. Clin Nutr. 28:39-45.

[41] Giacco R, Clemente G, busiello L, Lasorella G, Rivieccio AM, Rivellese AA, Riccardi G (2004) Insulin sensitivity is increased and fat oxidation after a high-fat meal is reduced in normal-weight healthy men with strong familial predisposition to overweight. Int J Obes. 28(2):342-8.

[42] Tentolouris N, Alexiadou K, Kokkinos A, Koukou E, Perrea D, Kyriaki D, Katsilambros N (2011) Meal-induced thermogenesis and macronutrient oxidation in lean and obese women after consumption of carbohydrate-rich and fat-rich meals. Nutrition. 27(3):310-315.

[43] Coelho SB, Sales RL, Iyer SS, Bressan J, Costa NMB, Lokko P, Mattes R (2006) Effects of peanut oil load on energy expenditure, body composition, lipid profile, and appetite in lean and overweight adults. Nutrition. 22:585-592.

[44] Rasmussen LG, Larsen TM, Mortensen PK, Due A, Astrup A (2007) Effect on 24-h energy expenditure of a moderate-fat diet high in monounsaturated fatty acids compared with that of a low-fat, carbohydrate-rich diet: a 6-mo controlled dietary intervention trial. Am J Clin Nutr. 85:1014-1022.

[45] St-Onge M-P, Ross R, Parsons WD, Jones PJH (2003) Medium-chain triglycerides increase energy expenditure and decrease adiposity in overweight men. Obes Res. 11(3):395-402.

[46] Kratz M, Callahan HS, Yang PY, Matthys CC, Weigle DS (2009) Dietary n-3-polyunsaturated fatty acids and energy balance in overweight or moderately obese men and women: a randomized controlled trial. Nutr Metab. 6:24.

Applications of Calorimetry into Propellants, Alloys, Mixed Oxides and Lipids

Numerical Solutions for Structural Relaxation of Amorphous Alloys Studied by Activation Energy Spectrum Model

Kazu-masa Yamada

Additional information is available at the end of the chapter

1. Introduction

What is the physical process which is producing the general characteristic of glass? It is said that the transformation of the liquid-system to glass one is a final theme in physics through into the twenty-first century. Furthermore, the development of amorphous-material devices and specimen modification methods is closely related with the obviousness of high thermal stability and stability of relaxation.

The aim of this research is to clarify numerical solutions for nano-structure relaxation processes focusing on the activation energy in transition metal (ie Cu, Fe) based amorphous alloys. Activation energy for structural relaxation process in a metal type amorphous ternary and quaternary alloys, with cross sections of typically 0.03 mm x 2.0 mm, prepared by chill-block melt spinning has been investigated by Differential Scanning Calorimetry (DSC) with a cyclically heating technique [1,2,3]. Activation energies for structural relaxation with a spatial quantity in amorphous materials have been discussed by use of a total relaxed ratio function that depends on annealing temperature and time. In the present work in amorphous ternary and quaternary alloys, the distributions for the Activation Energy Spectrum (AES) with derivative-type relaxed ratio function were observed. Another result has been also established that the "reversible" AES model energy distribution though the cyclically nano-structural relaxations were in good agreement with the presented experimental results of transition metal based amorphous alloys.

There has been recently considerable that the glassy alloys are representative of the bulk formed ultra-fine structure [1]. Particularly Cu has been shown to be good base element for bulk glass-forming alloy with fully glassy sections recently by use of die injection casting [2,3]. Binary Cu - (Zr or Hf) alloys have been found to form an amorphous phase over a

108 Calorimetry: Modern Concepts and Applications

wide composition range. However, addition of Ti in both these binary systems greatly increased the glass forming ability (GFA), with the critical diameter for fully amorphous rods being at least 4 mm for $Cu_{60}Zr_{30}Ti_{10}$, $Cu_{60}Hf_{20}Ti_{20}$ and $Cu_{55}Hf_{25}Ti_{20}$ [2,3]. Meanwhile the understanding of the structural relaxation process is essential in the development of stability for amorphous alloys, as well as in establishing stable working temperature to avoid the degradation of strength. Therefore, high thermal stability of quasi-stable amorphous materials for Cu based alloys. The atomic mechanism of diffusion in amorphous alloys is still poorly understood as compared to that in crystalline alloys. However, measurements of diffusivity in amorphous alloys have been limited so far because of the experimental difficulties of measuring the very small diffusion coefficients, usually less than 10^{-17} m^2s^{-1}, which are typical of amorphous alloys below their crystallization temperatures [4,5].

In the present work, using Differential Scanning Calorimetry (DSC) thermal analysis has been made to determine the activation processes [6,7,8], and to evaluate whether it represents the thermodynamically stable form of CuHfTi and CuHfTi-B glass-forming amorphous alloys.

2. Experimental procedure

Cu-based alloy ingots of composition $Cu_{60}Hf_{20}Ti_{20}$, $(Cu_{60}Hf_{22}Ti_{18})_{0.99}B_1$ and $(Cu_{60}Hf_{22}Ti_{18})_{0.97}B_3$ were prepared by arc-melting mixtures in an argon gas atmosphere purified with a Ti getter. The alloy compositions represent the nominal values but the weight losses in melting were negligible. The alloy ingots were inverted on the hearth and re-melted several times, to ensure compositional homogeneity. Ribbon samples of each alloy, with cross sections of typically 0.03 mm * 2.0 mm, were produced by chill-block melt spinning in a sealed inactive gas atmosphere. The amorphous state of the specimen of the ribbon samples was confirmed by X-ray diffraction.

The endothermic/exothermic heats for relaxation process were measured by differential scanning calorimetry (DSC) of DSC3100s of MacScience Co., Ltd (Bruker Japan Co., Ltd.) at a constant heating rate of 1.00 K/s. And the ordered specimens were prepared by annealing used in the electric furnace of the DSC. Pre-annealing and following long-time main annealing by a DSC furnace are at a 700 K for 1800 s and at a 580 K for 6000 s, respectively. The maximum temperature of 700 K is enough to suppress the crystallization and to measure optimistically the structural relaxation for these three kinds of specimen [9].

2.1. Theory with DSC annealing process

Due to insufficient data of thermal stability of amorphous alloys, the following points are left as future problems. Even bulk glass-forming alloy, also amorphous alloys is a non-equilibrium state. The certain overall atoms in an amorphous alloy are in non-stable state rather than in the stable crystalline state. Therefore, not only crystallization over a certain wide temperature range but also re-arrangement of atoms occurs. The structural relaxation process is one of the essential phenomena in some non-equilibrium materials. Thereby, to study the structural relaxation is important to investigate the constitutional property of the amorphous alloys. Furthermore, the structural relaxation is closely connected with the stability of specific examples related to the bulk glass-forming amorphous alloy. It is also necessary to know this property from a viewpoint of the application development.

Consider the population of an assembly of reaction centre for structural relaxation, that is to say isolated double wells potential model (or so called Two Level System, TLS) as shown in Fig. 1[9]. A relaxation centre which is isolated and in a particular structural configuration, permits an atom to be either in a higher energy position at state 0 or in a lower energy position at state 1 in Fig. 1. The axis of abscissas is the configuration variable for relaxation processes, and the position 1/2 on the axis in Fig.1 is the saddle point for the energy wall between the position 0 and 1.

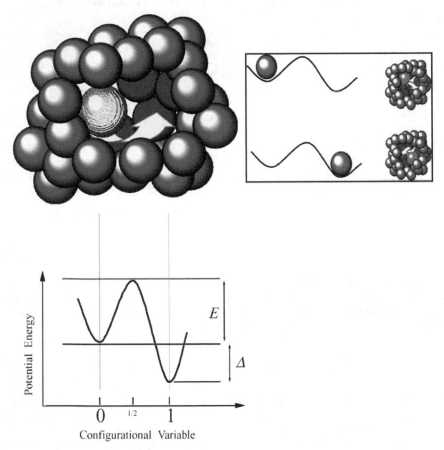

Figure 1. Schematic illustrations of relaxation centre and energy levels for the corresponding two level system

On the population of an assembly of this model, activation energy spectrum (AES) in structural relaxation processes, J. A. Leake, J. E. Evetts and M. R. J. Gibbs [10,11] describe the phenomena of physical available and variable property with good agreement between the theory and the experimentation. The theory assumes exponent factor nearly equal one-dimension for chemical reaction kinetics of Jhonson-Mehl-Avrami (JMA) equation.

Therefore, the theoretical model for the relaxation process in amorphous materials on the basis of a spectrum of available processes with a distribution of activation energy was proposed. In their model, the total change in the measured property, ΔP is given by

$$\Delta P = \int_0^E p(E)dE \tag{1}$$

In the range of activation energy E to $E+dE$ during the structural relaxation process, the total available property $p_0(E)$ changes such as

$$p(E)dE = p_0(E)\left[1-\exp\left\{-v_0 t \exp\left(-\frac{E}{kT}\right)\right\}\right]dE \tag{2}$$

where v_0 is an order of the Debye frequency ($v_0 \cong 10^{12}$ Hz) . Primak [12] rewrites Eq. (2) as

$$p(E) = p_0(E)\,\theta\left(E,T,t\right), \tag{3}$$

where $\theta(E, T, t)$ is defined as the characteristic annealing function. Thus, the function of $\theta(E, T, t)$ is a measure of the proportion of available processes at the energy E. Proportion as $\theta(E, T, t)$ has contributed to the relaxation property after the time t at the annealing temperature T. The form of $\theta(E, T, t)$ is given in Fig. 2 (a) and (b) .

In the another paper [9], in a process for most simplifying assumption, the function $\theta(E, T, t)$ can be replaced by step function at an energy $E_0(T, t)$ as

$$E_0 = kT \ln\left(v_0 t\right) \tag{4}$$

These E_0 forms are given in Fig. 2 (c, d, e, f, g, h, i), then E_0 changes from 0 to 1 over one-step meanwhile θ (E) changes over a narrow range of E and T. If in the simplified calculations, the E_0 should have been applied. On the other hand, in the present work for calculation using specific 1st derivative-type relaxation ratio, as follows.

$$\frac{d\theta\left(E,T,t\right)}{dE} = \left\{v_0 t \exp\left(-\frac{E}{kT}\right)\right\}\left(-\frac{E}{kT}\right)\exp\left\{-v_0 t \exp\left(-\frac{E}{kT}\right)\right\} \tag{5}$$

The (5) function is shown in Fig 3 (a), on the contrary the derivative of $E_0(T)$ is shown in Fig 3 (b) . So it should be preferred replacement with a better approximation using the function (5). By using simple area summation method, furthermore the normalization for the linear function of S=0.234 T+0.244 in this case as shown in Fig 4 would be applied, we can estimate the actual activation energy spectra distributions using the method of Fig 5. It is so called normalized 1st derivative - type relaxation ratio function in our organized work.

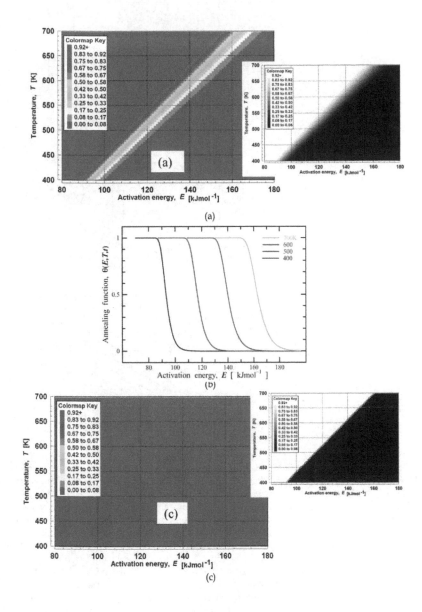

(a)

(b)

(c)

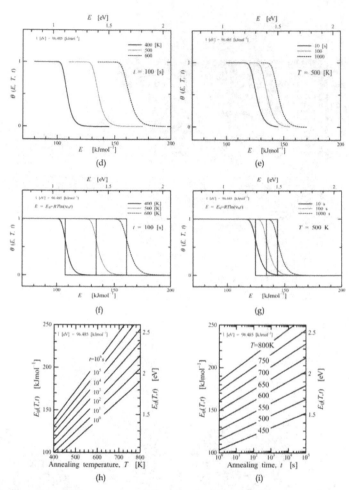

Figure 2. (a). Dependence of the characteristic annealing function $\theta(E, T, t=1s)$ on the activation energy E and the temperature T [9,10]. Fig. 2(a) is used of $\theta(E, T, t=1s)$ (b). Dependence of the characteristic annealing function $\theta(E, T, t=1s)$ on the activation energy E and the temperature T [9,10]. Fig. 2(b) is used of $\theta(E, T, t=1s)$ (c). Dependence of the characteristic annealing function $\theta(E, T, t=1s)$ on the activation energy E and the temperature T [9,10]. Fig. 2(c) is used of $E_0(T, t=1s)$. Right hand and left hand of charts are the same figure. Additionally right one is shown in grey scale for visible-contrasty (d,e,f,g). Dependence of the characteristic annealing function $\theta(E, T, t)$ and step function of threshold $E_0(T, t)$ on the horizontal axis of the activation energy E [9,10]. Fig. 2(d) at the upper left is used of $\theta(E, T, t=100s)$. Fig. 2(e) at the upper right is used of $\theta(E, T=500K, t)$. Fig. 2(f) at the lower left is used with step functions superposed on $\theta(E, T, t=100s)$. Fig. 2(g) at the lower right is used with step functions superposed on $\theta(E, T=500K, t)$ (h, i). Dependence of the simplifying assumption of characteristic annealing function, activation energy $E_0(T, t)$ vs the annealing time t and the isothermal annealing temperature T [9,10]. Fig. 2(h) at the left is used of $E_0(T, t=1, 10, 10^2, 10^3, 10^4, 10^5$ and 10^6 s). Fig. 2(i) at the right is used of $E_0(T=450, 500, 550, 600, 650, 700, 750$ and $800K)$

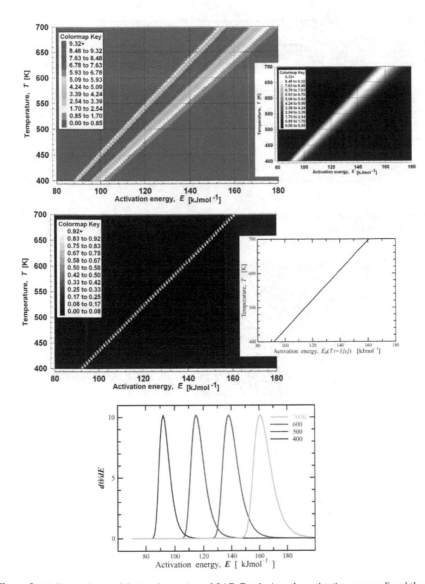

Figure 3. (a). Dependence of the 1st derivation of θ (E, T, t=1 s) on the activation energy E and the temperature T. Fig. 3 (a) Derivative type θ(E, T, t=1 s) . Right hand and left hand of charts are the same figure. Additionally right one is shown in grey scale for visible-contrasty (b). Dependence of the 1st derivation of θ (E, T, t=1 s) on the activation energy E and the temperature T. Fig. 3 (b) Approximation type $E_0(T, t$=1 s). Right hand and left hand of charts are the same figure. Additionally right one is for visible-contrasty (c). Dependence of the 1st derivation of θ (E, T=400, 500, 600 and 700K, t=1 s) on the activation energy E.

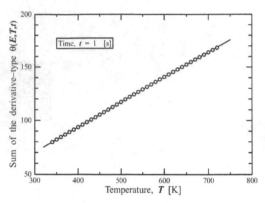

Figure 4. Dependence of the summation S of $1st$ derivative type $\theta(E, T, t=1s)$ on the temperature T. Plots can be replaced by linear function with a good approximation in $S = 0.234\ T + 0.244$

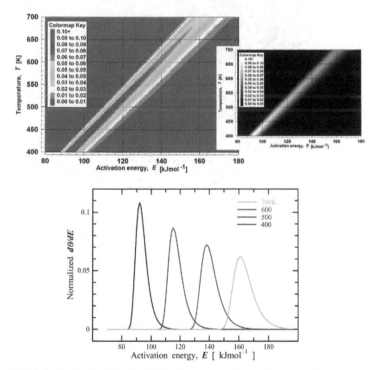

Figure 5. (a). Divided by the $S = 0.234\ T + 0.244$ of Fig 4, dependence of the normalized 1st derivative - type relaxation ratio function of $\theta(E, T, t=1s)$ on the activation energy E and the temperature T. Right hand and left hand of charts are the same figure. Additionally right one is shown in grey scale for visible-contrasty (b). Divided by the $S = 0.234\ T + 0.244$ of Fig 4, dependence of the normalized 1st derivative - type relaxation ratio function of $\theta(E, T=400, 500, 600$ and $700K, t=1$ s) on the activation energy E

2.2. Experimental processes with DSC annealing

In the previous paper [9], we discussed at first the AES applied on the Cu-Hf-Ti system,
because of the following reasons.

In Table 1 [2] the glass forming ability (GFA) related to the T_{rg} ($=T_g / T_L$) was observed that
addition of Ti in Cu-Hf binary systems greatly increased. In this table, they were shown
with the critical diameter dC for fully amorphous rods being at least 3 mm for
$Cu_{60}Hf_{22.5}Ti_{17.5}$, $Cu_{60}Hf_{20}Ti_{20}$ and $Cu_{60}Hf_{17.5}Ti_{22.5}$.

Composition	dc	T_g (K)	T_x (K)	T_L (K)	T_{rg} ($=T_g / T_L$)
$Cu_{60}Hf_{35}Ti_5$	Ribbon	775	828	1333	0.58
$Cu_{60}Hf_{30}Ti_{10}$	Ribbon	751	806	1263	0.59
$Cu_{60}Hf_{25}Ti_{15}$	Ribbon	748	788	1240	0.60
$Cu_{60}Hf_{22.5}Ti_{17.5}$	4mm	745	780	1234	0.60
$Cu_{60}Hf_{20}Ti_{20}$	4mm	740	767	1211	0.61
$Cu_{60}Hf_{17.5}Ti_{22.5}$	3mm	732	755	1229	0.59
$Cu_{60}Hf_{15}Ti_{25}$	Ribbon	722	745	1223	0.59
$Cu_{60}Hf_{10}Ti_{30}$	Ribbon	700	726	1233	0.56
$Cu_{60}Hf_5Ti_{35}$	Ribbon	690	712	1223	0.56

Table 1. Glass forming section thickness and thermal property for $Cu_{60}Hf_{40-x}Ti_x$ (X=from 5 to 35) alloy
series [2]

After all, the aim of this research is also to clarify a quantitative evaluation in the structure
relaxation processes focusing on the activation energy in $Cu_{60}Hf_{20}Ti_{20}$ based amorphous
alloys with high GFA series.

Pre-annealing and main-annealing conditions were completely similar the way as Ref. 9.
After that it will be noted that an atom to be in a higher energy position at stage 0 in Fig. 1,
endothermic heat occurs at 1st run in the measurement scanning #1 even at 1st run in #2,
that is to say the reversible relaxation processes, and the 1st minus 2nd run indicates the
endothermic value (the 2nd minus 1st run indicates the exothermic value) that could be
calculate the AES distributions that means an atom to be in a higher energy position at stage
0 in Fig. 1.

The liquidus temperature T_L has its minimum value for the 20at%Ti alloy in Table 1,
probably corresponding to a ternary eutectic system, it is because only this composition has
single melting peak in the DTA trace [2]. Thus this is a dominating factor in determining
that T_{rg} has its maximum value at this composition clearly. The atomic diameter of Ti
(0.289nm) is intermediate between those of Cu (0.256nm) and Hf (0.315nm) and, evidently,
equal proportions of Hf and Ti result in maximum stabilization of the densely packed liquid
structure [2] and normalized 1st derivative - type relaxation ratio function.

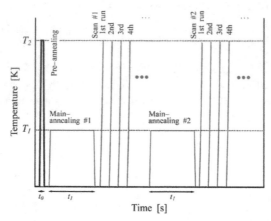

Figure 6. An example of annealing time and temperature history by use in fully electric furnace of the DSC. Specimens were prepared by annealing used in the DSC furnace, pre-annealing and following long-time main-annealing are at 700 K (=T_2) for 1800 s (=t_0) and recycled at a 580 K(=T_1) for 6000 s(=t_1), respectively. The maximum temperature of T_2 is enough to suppress the crystallization and to measure optimistically the structural relaxation following the continuous scan #1 1st run to n-th run. The "reversible" phenomena have been observed in the anneal process of scan #2 to scan #n as shown in Fig.7.

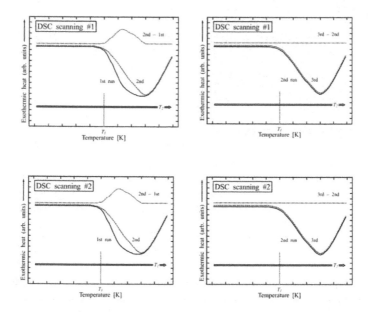

Figure 7. four-leafed schematic illustrations of reversible phenomena for DSC scanning #1 then #2, and included relaxation processes for scanning 1st run then 2nd run, on the other hand without relaxation processes for scanning 2nd run then 3rd run

3. Results through experimental procedure

In the presented work by use of the AES model, following above-mentioned, activation energies in structural relaxation processes have been determined of composition $Cu_{60}Hf_{20}Ti_{20}$ and related $(Cu_{60}Hf_{22}Ti_{18})_{0.99}B_1$ and $(Cu_{60}Hf_{22}Ti_{18})_{0.97}B_3$ amorphous alloys as shown in Fig. 8.

The maximum energies in AES have similar tendency among three kinds of alloy nearly at 160 kJmol^{-1} (1.66 eV). Between the three kinds of B for 3, 1 and 0 % alloys, in an energy region less than 160 kJmol^{-1}, AES of only B 3 % alloy is higher than that of B 1% and 0 %. Meanwhile, in an energy region more than 160 kJmol^{-1}, AES of them are similar.

This suggests that the diffusion path size for the diffusant of Ti that atomic radius is smallest in the metallic compositions and the packing density of the covalent bonding matrix between the boron and metal are dominant in the relaxation processes. Consequently activation energy for the structural relaxation process has been determined in the $Cu_{60}Hf_{20}Ti_{20}$ with having the highest bulk glass-forming ability in $Cu_{60}Hf_{40-x}Ti_x$ (X are from 5 to 35 %) alloy series as almost 160 kJmol^{-1} (1.66 eV) using the normalized derivative - type relaxed ratio function [9].

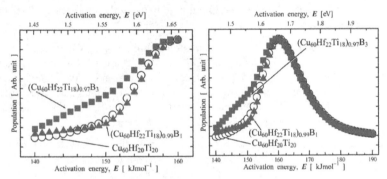

Figure 8. *In the present work for calculation using specific* normalized *1st derivative - type relaxation ratio function of* $\theta(E, T, t=1s)$ *, activation energy spectrum distributed in* $Cu_{60}Hf_{20}Ti_{20}$, $(Cu_{60}Hf_{22}Ti_{18})_{0.99}B_1$ and $(Cu_{60}Hf_{22}Ti_{18})_{0.97}B_3$

4. Calculation technique

In the Fig. 7., four-leafed schematic illustrations show on the DSC scanning #1, #2, we could evaluate the value included relaxation processes for scanning 1st run then 2nd run, on the other hand we could estimate the value included not relaxation processes for scanning 2nd run then 3rd run that are almost without relaxation. So solving the value should be calculated by DSC exothermic heats of 2nd run minus 1st run. So it is very important to calculate the differences between 2nd and 1st run. But also it is difficult to calculated form the DSC exothermic heats of 2nd run minus 1st run because of the temperature scanning step problem. This section describes the way of calculation how to get the differences (distinction) data.

In the Fig. 9., if they were a typical numerical example for differential calculation with supplied as text-type file name, for example, TEST00.TXT of 1th DSC run and TEST01.TXT of 2nd one, it would be transformed from their differential calculation to result numerical data such as TEST04.TXT, to be free to use a program such as GP.EXE ver. 4.13 and DOSBox version 0.74. The 2-Dimension Graph Plotter GP.EXE version 4.13-PC/AT and Dos-emulator DOSBox version 0.74 for all kind of MS-windows OS (another DOSBox version exists for MAC OS probably) are both free software supported in English keyboard peripheral interface. Additionally information, the GP.EXE was built by Prof. Dr. K. Edamatsu(now at riec.tohoku.ac.jp) in 1980-99 year for design to plot scientific/engineering graphs using PCs. Prof. Edamatsu said in his GP's documentation "with GP.EXE, make smart graphs for your presentation and publication. Also, try GP's powerful data analysis capability such as general least-squares fitting, numerical differentiation and integration".

In around 2010 year, the GP.EXE with super high speed and powerful data analysis capability is born-again by use of high performance Dos-emulator DOSBox.

Presented process, to calculate the differential exo/endothermic heat supplied from DSC live scanning environmental with gas flow atmosphere. Overall, Relaxation processes for example has been tutorial as bellow description mainly using the freeware GP.EXE, further only using scanning data 1st run to 2nd run.

In the Fig. 10., for introduction to present calculation technique, typical complex 2-D tutorial graph samples are shown by using GP.EXE. A left chart is the typical Gaussian differentiation tutorial sample, 1st derivative and 2nd one and experimental data and calculation. A right chart is the typical Ahhrenius tutorial plot with inversed horizontal axis with logarithm vertical axis. If you were to use the GP.EXE, you should download from the site of www.vector.co.jp/soft/dos/business/se004831.html.

Then you could get the file of gpat431.lzh, you should make the directory for set the GP.EXE environments. It should be save and destination to (recommended): C:/prog/gp/gp.exe, C:/prog/gp/INIT.GPR, C:/prog/gp/DOC, C:/prog/gp/ DRIVERS, etc.

Note: GP.EXE system is so called legacy-DOS, overall generated user filename must be kept the name rule of 8 character letter filename and 3 character letter extensions around in the GP directory.

Addition you could get the file of gpsmp420.lzh of tutorial examples of GPR extension files, you could be easy to get the way the GP.EXE operating. As shown in Fig. 10., the tutorials exist in site of www.vector.co.jp/soft/dos/business/se010753.html.

In the Table 2. , Recommended Dos-emulator DOSBox version 0.74 configuration file descriptions are shown. You could edit (ie. MS-Win7) it in Program Menu, DOSBox options, editing the configuration, last lines for "autoexec" region. Addition, keyb command needs the user of Japanese JP106 keyboard peripheral interface only (addition the keyb program and keyboard-map should also be needed. The keyb system's useful information would be gathered in World Wide Web). Meanwhile for in English peripheral US101 user, the keyb command should be ignore. Furthermore the last gp command in table 2, it should not be need to user for non-automatic start of GP. For normal user, it should be ignore the last one. Otherwise all users command the type key of gp on DOSBox command line, GP.EXE starts anytime.

```
[autoexec]
# Lines in this section will be run at
startup.
# You can put your MOUNT lines here.
@ECHO OFF
MOUNT c C:\
c:
```

Table 2. A sample of Dos-emulator DOSBox version 0.74 configuration file description (as shown in autoexec area only)

In the Fig. 11., GP.EXE column structure menu indicating live date column was shown. In the case, X, Y, YE of default column structure allow to use a delimiter also space and tabulator key. Live data should be minimum structure of X and Y with delimiter of space key. Meanwhile additional data column if include could be were specially galloped by use of "U" rule for GP column structure. The typical sample structure of live data is shown in Table 3a and 3b. Addition the 1st, 2nd and 3rd line were normally (default) galloped through a whole text-file for GP because of a purpose for a title and axis captions.

INIT.GPR and other GPR files would be able to modified by a text-type general-purpose editor, then directory file path, captions and so on in them could be also re-arranged and rapid setting for similar graph format preparations.

Note: GPR file always includes full-Path towards live data, but usually it is NOT often need to full-Path towards them but only Local-Path that means without non-Path description, then some of this full-Path should be deleted by use of a text-type general-purpose editor because of keeping the safety-connection between the live data and the GPR file.

In the Fig. 12., load file name menu indicates the live date formatted general-purpose text-style pursuant to table 3a,3b. As it was shown, 3 files (2 kinds of file) of TEST00.TXT, TEST01.TXT and TEST00.TXT are loaded in the live data tray in GP.EXE for 2 data differential calculations. The file #1 should be without differential calculations. The file #2 and #3 should be with differential calculation for #3 minus #2.

In the Fig. 13., load and save parameter's file (GPR of extensions) menu indicates graph structure list organized whole graphic design. Especially GPR file is also plain text-type, so we could arrange them before/afterward by use of a text-type general-purpose editor anytime.

Note: GP.EXE system is so called legacy-DOS, overall generated user filename must be kept the name rule of 8 character letter filename and 3 character letter extensions around in the GP directory.

In the Fig. 14., a Interfile Calculation Parameters and style-menu displayed, red-mark of TEST01 (Src. file # 2) on left-y-axis and white-mark of TEST00 (Src. file # 1) on left-y-axis, meanwhile blue-mark of TEST00 minus TEST01 (Src. file #3 minus #2) on right-y-axis. The aim of this computation is to process the file #3 minus #2 based rule on column X date for Temperature region. Finally a result of the processed data has been shown as blue-mark beside on right-y-axis.

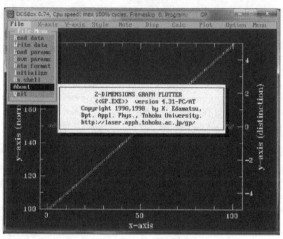

Figure 9. 2-Dimension Graph Plotter GP.EXE version 4.13-PC/AT and Dos-emulator DOSBox version 0.74 are both free software in English supported to calculate the differential exothermic heat data using DSC included relaxation processes for example, scanning 1st run to 2nd run.

GP.EXE: http://www.vector.co.jp/soft/dos/business/se004831.html
GP.EXE samples :http://www.vector.co.jp/soft/dos/business/se010753.html

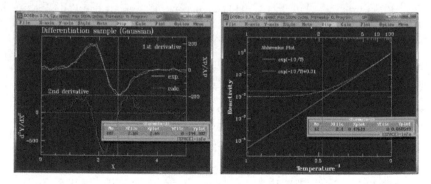

Figure 10. Typical samples of 2-Dimension Graph Plotter GP.EXE. A left chart is the typical Gaussian differentiation sample, 1st derivative and 2nd one and experimental data and calculation. A right chart is the typical Ahhrenius-type plot with inversed horizontal axis together with logarithm vertical axis. Green colour cross line indicator means the across point both live-data and translated-data.

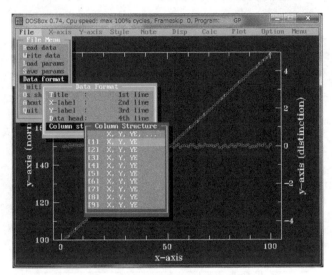

Figure 11. At first, column structure menu indicates the live date column structure

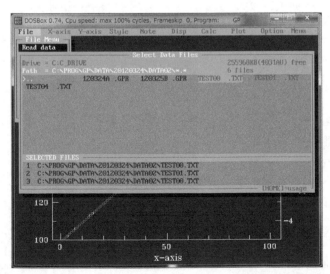

Figure 12. Second, load file name menu indicates the live date formatted general-purpose text-style
pursuant to table 3a,3b

x-axis-data	y- axis-data
A main title of presented Graph	
A title of x-axis	
A title of y-axis	
1.022326732	100
2.010754735	101
2.943458995	102
4.094871558	103
4.993183668	104
6.073690076	105
7.054419829	106
•	•
•	•
•	•
97.94832211	197
98.91335378	198
100.0896175	199

a

x-axis-data	y- axis-data
A main title of presented Graph	
A title of x-axis	
A title of y-axis	
1.00	100
2.00	101
3.00	102
4.00	103
5.00	104
6.00	105
7.00	106
•	•
•	•
•	•
98.00	197
99.00	198
100.00	199

b

Table 3. a. Typical numerical example for differential calculation with random number generator only onside x-axis formatted for GP.exe as data filename TEST01.TXT b. Typical numerical example for differential calculation formatted for GP.exe as data filename TEST00.TXT

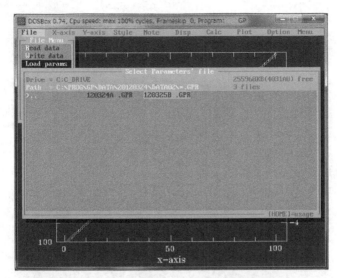

Figure 13. Third, load parameter's file menu indicates organized graph structure

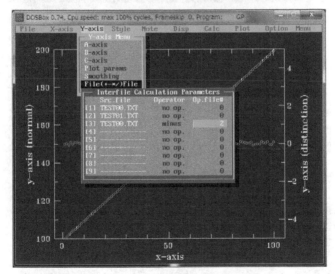

Figure 14. Forth, *an Interfile Calculation Parameters and* style-menu displayed, red-mark of TEST01 (Src. file # 2) on left-y-axis and white-mark of TEST00 (Src. file # 1) on left-y-axis, meanwhile blue-mark of TEST00 minus TEST01 (Src. file #3 minus #2) on right-y-axis. In the figure, a highlight area of green, the letter of the 2 means the #2 file that selected for Src. file #3 minus #2.

In the Fig. 15., Left and right axis, so called Y-axis Plotting Parameters are shown relation to Fig.14. For Src. file #1 and 2 are to belong to left-y-axis named A of Y-axis and further calculated Src. file #3 minus #2 is to belong to right-y-axis named B.

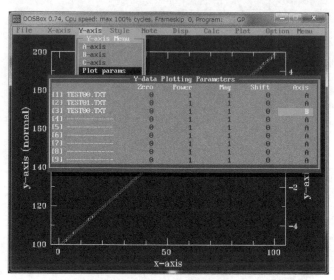

Figure 15. Fifth, Y-axis Plotting Parameters are shown. For Src. file #1 and 2 are to belong to left-y-axis named A and Src. file #3 minus #2 (calculated data) is to belong to right-y-axis named B

In the Fig. 16., Plotting green cross-line indicator means the calculated Src. file #3 minus #2 dots. Fig. 10s are also the similar for usage of cross-line indicator.

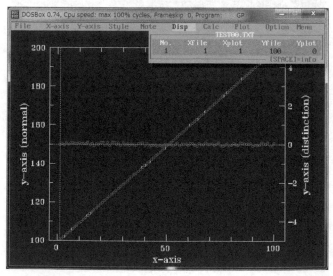

Figure 16. Sixth, Plotting green cross-line indicator means the calculated Src. file #3 minus #2 dots

In the Fig. 17., It is the most important method for calculating of relaxation process. Text-fire save-menu using blue-mark of TEST00 minus TEST01 (Src. file #3 minus #2) on right-y-axis should be describe for example as file name TEST04.TXT in write-data filename input region. Then the TEST04.TXT should be on further calculation process to equation (5) , $t=1(s)$, then should be normalized and multiplied by inverse reactor summation.

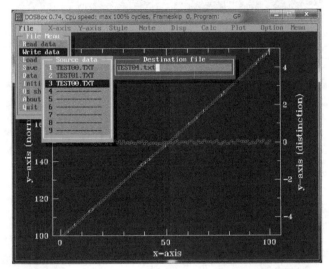

Figure 17. Seventh, text-fire save-menu using blue-mark of TEST00 minus TEST01 (Src. file #3 minus #2) on right-y-axis as file name TEST04.TXT displayed in write-data panel

In the Fig. 18., PostScript-file save-menu using a PostScript-file driver of PS.DLL, that include gpat431.lzh archive, displayed in Plot Parameters panel. Furthermore the useful information, if it assumed to be a 01.ps as saved file name for presented graph design, it would be transformed from PostScript-file to PDF-file, for example, from 01.ps to assumed 01gw.pdf, to be free to use a program ghostscript ver. 9.04. It should be typed on command-line supported by each OS in current directory of 01.ps (not use the command-line in DOSBox) as:

```
"C:\Program Files\gs\gs9.04\bin\gswin32c.exe" -dNOPAUSE -dBATCH -
sDEVICE = pdfwrite -r600 -sOutputFile = 01gw.pdf -c 300000
setvmthreshold save pop -f 01.ps
```

Assumed 01gw.pdf would be a graph with super-resolution quality attaching suitable for all kind of publications. For example, it could be transformed from their PDF to word-processor MS-Word, to be free to use a program such as "Acrobat Reader", and it should be typing keys of Control-a, then Zoom up to around 200%, then Control-c, after then in word-processor to be also typing keys Control-v for universal use.

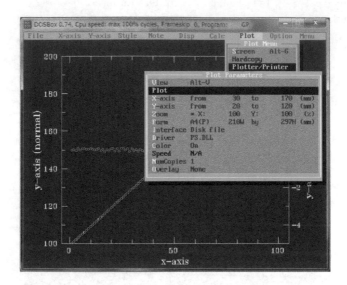

Figure 18. Final, PostScript-file save-menu using a PS.DLL PostScript-file driver displayed in Plot
Parameters panel. If it assumed to be a 01.ps as saved file name, it would be transformed from
PostScript-fire to PDF-file, for example, from 01.ps to assumed 01gw.pdf, to be free to use a program
ghostscript ver. 9.04. It should be typed on command-line as:
```
"C:\Program Files\gs\gs9.04\bin\gswin32c.exe" -dNOPAUSE -dBATCH -sDEVICE=pdfwrite -r600
-sOutputFile=01gw.pdf -c 300000 setvmthreshold save pop -f 01.ps
```

5. Conclusion

In the present work for calculation using specific normalized 1st derivative - type relaxation
ratio function of $\theta(E, T, t=1s)$, activation energy spectrum distributed in $Cu_{60}Hf_{20}Ti_{20}$,
$(Cu_{60}Hf_{22}Ti_{18})_{0.99}B_1$ and $(Cu_{60}Hf_{22}Ti_{18})_{0.97}B_3$ have been observed through the process of
numerical-based discussion. It is so called rapid-type clarification between the temperature
range T_2 and T_1 for almost around narrow 100 K region. In other words, even the above
mentioned narrow temperature range induced the "reversible" phenomena, and it has been
also observed in the anneal process of scan #1 to #n, repeatedly.

After it has been difficult in general to calculate numerical differences between any kinds of
DSC live data. Because it has the time-domain problem for stepping accuracy and speed on
temperature column region. So in second half of this paper, it was tutorial to short course
calculation method for the differences using the freeware in Tohoku University Prof. K.

Edamatsu' GP.EXE that was designed until 1999 to make smart graphs for publication with powerful data analysis ability such as numerical complex differentiation. And now it is shown that the GP.EXE has been useful for genuine data processing even in the 2012's generation.

Author details

Kazu-masa Yamada
Hakodate National College of Technology,
Department of Electrical and Electronic Engineering,
Japan

Acknowledgement

The author was favoured to have the assistance of Dr. I. A. Figueroa in Universidad Nacional Autonoma de Mexico who contributed an experimental circumstance to the accomplishment of the amorphous sample preparations in the University of Sheffield UK. The author also would like to express the appreciation to Dr. Sergio Gonzalez Sanchez (Universitat Autonoma de Barcelona), Mr. P. J. J. Hawksworth and Dr. I. Todd in the University of Sheffield. The author is indebted to Professor H. A. Davies for drawing his attention to presented researches.

6. References

[1] Brochure for the Nanotechnology and Materials Technology Development Department, on March (2008), The New Energy and Industrial Technology Development Organization (NEDO) in Japan, processing technology for metallic glasses, p.61-62, http:// www.nedo.go.jp/ kankobutsu/ pamphlets/ nano/ nano_e2008.pdf

[2] I.A. Figueroa, R. Rawal, P. Stewart, P.A. Carroll, H.A. Davies, H. Jones and I. Todd, Journal of Non-Crystalline Solids, Vol. 353 (2007), pp. 839-841

[3] I.A. Figueroa, H.A. Davies, I. Todd and K. Yamada, Advanced Engineering Materials, Vol. 9 (2007), pp. 496-499

[4] K. Yamada, Y. Iijima and K. Fukamichi: Defect and Diffusion Forum, Vols. 143-147 (1997), pp.765-770

[5] K. Yamada, Y. Iijima and K. Fukamichi: J. Mater. Res., Vol. 8 (1993), pp.2231-2238

[6] Y. Takahara, A. Morita, T. Takeda and H. Matsuda: J. Japan Inst. Metals, Vol. 54 (1990) pp. 752-757 (in Japanese)

[7] K. Yamada, M. Ito, M. Tatsumiya, Y. Iijima and K. Fukamichi: Defect and Diffusion Forum, Vols. 194-199 (2001), pp.815-820

[8] K. Yamada, K. Fukamichi and Y. Iijima: J. Magn. Soc. Jpn., Vol.22 Suppl. S2 (1998), pp. 97-100

[9] K. Yamada et al, Defect and Diffusion Forum Vols. 283-286 (2009), pp 533-538

[10] J. A. Leake, E. Woldt, and J. E. Evetts, Mater. Sci. Eng. , Vol. 97 (1988), pp. 469-472
[11] M. R. J. Gibbs, J. E. Evetts and J. A. Leake, J. Mater. Sci., Vol. 18 (1983), pp. 278-288
[12] W. Primak: Phys. Rev. Vol. 100 (1955) pp. 1677-1689

Thermal Decomposition Kinetics of Aged Solid Propellant Based on Ammonium Perchlorate – AP/HTPB Binder

R. F. B. Gonçalves, J. A. F. F. Rocco and K. Iha

Additional information is available at the end of the chapter

1. Introduction

Despite the widespread use and long investigative history of ammonium perchlorate(AP)-fuel mixtures, it still can be said that AP alone and AP/HTPB (hydroxyl-terminated-polybutadiene) composites remain among the most confounding materials in the research setting [1]. Since the physical structure of composite propellants like the AP/HTPB composite is heterogeneous, the combustion wave structure appears to be also heterogeneous. During the combustion, at the burning surface, the decomposed gases from the ammonium perchlorate particles and fuel binder (HTPB) are interdiffused and produce diffusion flame streams. Due this, the flame structure of AP composite propellants is complex and locally three-dimensional in shape.

Ammonium perchlorate (NH_4ClO_4) is a powerful oxidizer salt largely used in solid propellant formulations for application in airspace and defense materials industries. It is obtained by reaction between ammonia and perchloric acid, or by double decomposition between an ammonium salt and sodium perchlorate, and crystallizes with romboedric structure in room temperature and pressure, with relative density of 1.95 [2] Similarly to most ammonium salts, AP thermal decomposition occurs before its fusion. When submitted to a low heating rate, decomposes releasing gases chlorine, nitrogen and oxygen and water in the vapor state; while with a high heating rate stimulus there are instant reactions with high energy release.

During the combustion process of AP crystals at high pressures, is possible to observe the formation of a tiny layer of ammonium perchlorate in liquid phase at the grain surface [3], followed by a region where it is presented in gaseous phase.

According to Beckstead and Puduppakkam [4], the combustion of a monopropellant can be divided in three regions (condensed, liquid-gas two-phase region and gas region). The two-phase region consists of liquid and gaseous species resulting from the melting and/or decomposition of the solid phase. The precise division between the two-phase and gas-phase region (i.e. the 'burning surface') is not well defined due to chemical reactions, bubbles, and condensed material being convected away from the surface. In the gas phase region of a monopropellant, the flame is essentially premixed. The species emanating from the surface react with each other and/or decompose to form other species. A wide variety of reactions involving many species occur in the gas flame until equilibrium is reached in the final flame zone.

Thermal decomposition of AP, as its combustion processes, have been experimentally studied and reported in the literature. The thermal decomposition of AP may be observed by differential thermal analysis (DTA) and thermal gravimetry (TG), in the figure below [5]. A heating rate of 0.33 K/s was used on the analysis.

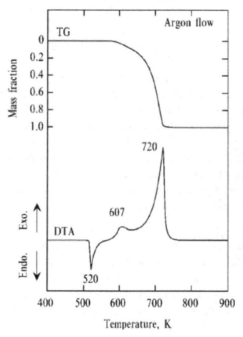

Figure 1. TG and DTA of AP decomposition [5]

The phase transition from orthorhombic to cubic crystal lattice (ΔH = -85 kJ/kg) is represented by the endothermic peak on 520 K. The exothermic events on 607 K and 720 K are due to the proper decomposition of the AP crystal in ammonia and perchloric acid, followed by the formation of chloridric acid and oxygen (decomposition of $HClO_4$), according to the reactions below.

$$NH_4ClO_4 \longrightarrow NH_3 + HClO_4$$
$$\downarrow$$
$$HCl$$
$$+$$
$$2 O_2$$

These oxygen molecules will be used as oxidizer in binders combustion, when the AP is used in a composite propellant or even when is burning by itself.

The combustion mechanism of AP has been studied and modified. The table below shows the elementary reactions which take part on the combustion process. This mechanism was proposed by Gross [6], according to literature data. It is very interesting the analysis of the combustion for its close relation to the thermal decomposition. When high pressure or high temperatures are used, the material suffers combustion instead of thermal decomposition, i.e. there's a higher velocity of decomposition and higher energy release, but the process is usually incomplete.

Reaction	A	b	Ea
$HClO_4=ClO_3+OH$	1.00E+14	0.0	3.91E+04
$HClO_4+HNO=ClO_3+H_2O+NO$	1.50E+13	0.0	6.00E+03
$ClO_3=ClO+O_2$	1.70E+13	0.5	0,00E+00
$Cl_2+O_2+M=ClO_2+Cl+M$	6.00E+08	0.0	1.12E+04
$ClO+NO=Cl+NO_2$	6.78E+12	0.0	3.11E+02
$ClO+ClOH=Cl_2+HO_2$	1.00E+11	0.0	1.00E+04
$ClOH+OH=ClO+H_2O$	1.80E+13	0.0	0,00E+00
$HCl+OH=Cl+H_2O$	5.00E+11	0.0	7.50E+02
$Cl_2+H=HCl+Cl$	8.40E+13	0.0	1.15E+03
$ClO+NH_3=ClOH+NH_2$	6.00E+11	0.5	6.40E+03
$NH_3+Cl=NH_2+HCl$	4.50E+11	0.5	1.00E+02
$NH_3+OH=NH_2+H2O$	5.00E+07	1.6	9.55E+02
$NH_2+O_2=HNO+OH$	3.00E+09	0.0	0,00E+00
$NH_2+NO=H_2O+N_2$	6.20E+15	-1.3	0,00E+00
$HNO+OH=NO+H_2O$	1.30E+07	1.9	-9.50E+02
$HNO+O_2=NO_2+OH$	1.50E+13	0.0	1.00E+04
$HNO+H=H_2+NO$	4.50E+11	0.7	6.60E+02
$NO+H+M=HNO+M$	8.90E+19	-1.3	7.40E+02
$HO_2+N_2=HNO+NO$	2.70E+10	0.5	4.18E+04
$NO+HO_2=NO_2+OH$	2.11E+12	0.0	4.80E+02
$H+NO_2=NO+OH$	3.47E+14	0.0	1.48E+03
$H_2+OH=H_2O+H$	2.16E+08	1.5	3.43E+03

Table 1. AP combustion mechanism

$k = A\ T^b \exp(-E/RT)$. Units: A (mol-cm-s-K), E (J/mol).

Based on combustion mechanisms, the burning process may be simulated and analyzed by some specific softwares. In a previous work [7], these simulations were done, considering a perfectly stirred reactor, different internal pressures and a specific temperature profile. The combustion simulation results may be observed in the figure below, which show the behavior of AP combustion with different internal pressures of the combustion chamber.

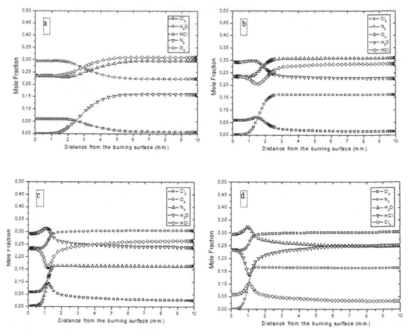

Figure 2. AP combustion at a) 1 atm; b) 5 atm; c) 30 atm and d) 60 atm.

The "elbows" appear due to the increase of the occurrence of intermediate reactions in the flame zone. This phenomenon generates a great variation on the mole fractions of intermediates (as the high temperature enhance the speed of the slower reactions, generating more radicals), which modify the concentration of the main species (specially in the flame zone), so the different slope is observed. As the pressure in the combustion chamber increases, there is an approximation of the flame to the material's surface and accentuation of the "elbows" presented on the flame region, indicating the influence of the speed increase of elementary reactions in the decomposition process of the material in study. This gain in chemical speed reactions may be converted in gain in thrust of rocket motors and specific impulse of solid propellant grains.

When a composite propellant is used, like AP-HTPB, the combustion process depends on the diffusion of the gases generated on the initial decomposition of the oxidizer, which surrounds the binder molecules at the burning surface. The combustion mechanism has higher complexity as new components are added, because there are the elementary reactions for each component decomposition and their interactions in chamber, as well as the formation and decomposition of new intermediary species, especially in the flame region. The proposed mechanism for AP-HTPB combustion may be observed in Table 2 below.

Reaction	A	b	Ea
$Cl_2+O_2+M=ClO_2+Cl+M$	6.00E+08	0	1.12E+04
$ClO+NO=Cl+NO_2$	6.78E+12	0	3.11E+02
$HCl+OH=Cl+H_2O$	5.00E+11	0	7.50E+02
$Cl_2+H=HCl+Cl$	8.40E+13	0	1.15E+03
$NH_3+Cl=NH_2+HCl$	4.50E+11	0.5	1.00E+02
$NH_3+OH=NH_2+H_2O$	5.00E+07	1.6	9.55E+02
$NH_2+O_2=HNO+OH$	3.00E+09	0	0.00E+00
$NH_2+NO=H_2O+N_2$	6.20E+15	-1.3	0.00E+00
$HNO+OH=NO+H_2O$	1.30E+07	1.9	-9.50E+02
$HNO+O_2=NO_2+OH$	1.50E+13	0	1.00E+04
$HNO+H=H_2+NO$	4.50E+11	0.7	6.60E+02
$NO+H+M=HNO+M$	8.90E+19	-1.3	7.40E+02
$HO_2+N_2=HNO+NO$	2.70E+10	0.5	4.18E+04
$NO+HO_2=NO_2+OH$	2.11E+12	0	4.80E+02
$H+NO_2=NO+OH$	3.47E+14	0	1.48E+03
$H_2+OH=H_2O+H$	2.16E+08	1.5	3.43E+03
$CH_4+Cl=CH_3+HCl$	2.50E+13	0	3.83E+03
$CH_4+H=CH_3+H_2$	6.60E+08	1.6	1.08E+04
$CH_4+OH=CH_3+H_2O$	1.00E+08	1.6	3.12E+03
$CH_3+H+M=CH_4+M$	1.27E+16	-0.6	3.83E+02
$CO+OH=CO_2+H$	4.76E+07	1.2	7.00E+01
$CO+ClO=CO_2+Cl$	3.00E+12	0	1.00E+03
$CO+ClO_2=CO_2+ClO$	1.00E+10	0	0.00E+00
$H+O_2=O+OH$	8.30E+13	0	1.44E+04
$CH_2+H_2=CH_3+H$	5.00E+05	2	7.23E+03
$CH_2+H+M=CH_3+M$	2.50E+16	-0.8	0.00E+00
$CH_4+O=CH_3+OH$	1.02E+09	1.5	6.00E+02
$OH+CH_3=CH_2+H_2O$	5.60E+07	1.6	5.42E+03
$C_2H_4+O_2=2CO+2H_2$	1.80E+14	0	3.55E+04
$NH_2+NO_2=2HNO$	1.40E+12	0	0.00E+00
$NH_2+ClO=HNO+HCl$	2.50E+12	0	0.00E+00
$O_2+HNO=NO+HO_2$	1.00E+13	0	1.30E+04
$H+Cl+M=HCl+M$	5.30E+21	-2	-2.00E+03
$Cl+Cl+M=Cl_2+M$	3.34E+14	0	-1.80E+03
$Cl+HO_2=ClO+OH$	2.47E+13	0	8.94E+02
$ClO+O=Cl+O_2$	6.60E+13	0	4.40E+02
$H+HCl=Cl+H_2$	7.94E+12	0	3.40E+03
$HCl+O=Cl+OH$	2.30E+11	0.6	9.00E+02

$Cl_2+O=Cl+ClO$	2.51E+12	0	2.72E+03
$N_2O+M=N_2+O+M$	6.20E+14	0	5.61E+04
$N_2O+OH=N_2+HO_2$	2.00E+12	0	2.11E+04
$N_2O+O=NO+NO$	2.90E+13	0	2.32E+04
$N_2O+O=N_2+O_2$	1.40E+12	0	1.08E+04
$N_2O+H=N_2+OH$	4.40E+14	0	1.89E+04
$2H+M<=>H_2+M$	1.00E+18	-1	0.00E+00
$2H+H_2<=>2H_2$	9.00E+16	-0.6	0.00E+00
$2H+H_2O<=>H_2+H_2O$	6.00E+19	-1.3	0.00E+00
$2H+CO_2<=>H_2+CO_2$	5.50E+20	-2	0.00E+00
$ClO_2+NO=ClO+NO_2$	1.00E+11	0	0.00E+00
$Cl+ClO_2=ClO+ClO$	5.00E+13	0	6.00E+03
$ClO+ClO=Cl_2+O_2$	1.00E+11	0	0.00E+00
$Cl+HO_2=HCl+O_2$	1.80E+13	0	0.00E+00
$Cl+O_2+M=ClO_2+M$	8.00E+06	0	5.20E+03
$NO_2+O=NO+O_2$	1.00E+13	0	6.00E+02
$HNO+HNO=H_2O+N_2O$	3.95E+12	0	5.00E+03
$NO_2+NO_2=NO+NO+O_2$	1.00E+14	0	2.50E+04
$Cl+N_2O=ClO+N_2$	1.20E+14	0	3.35E+04
$OH+OH=H_2O+O$	6.00E+08	1.3	0.00E+00
$NH_2+NO_2=H_2O+N_2O$	4.50E+11	0	0.00E+00
$HNO+NH_2=NH_3+NO$	5.00E+11	0.5	1.00E+03
$ClO+HNO=HCl+NO_2$	3.00E+12	0	0.00E+00
$HCl+HO_2=ClO+H2O$	3.00E+12	0	0.00E+00
$NH_2+NO=H+N_2+OH$	6.30E+19	-2.5	1.90E+03
$NH_2+OH=H_2O+NH$	4.00E+06	2	1.00E+03
$NH_2+NH_2=NH+NH_3$	5.00E+13	0	1.00E+04
$NH+NO=N_2+OH$	1.00E+13	0	0.00E+00
$NH+NO=H+N_2+O$	2.30E+13	0	0.00E+00
$Cl+NH_2=HCl+NH$	5.00E+10	0.5	0.00E+00
$ClO_2+NH=ClO+HNO$	1.00E+14	0	0.00E+00
$N+NO_2=NO+NO$	1.00E+14	0	0.00E+00
$N+N_2O=N_2+NO$	5.00E+13	0	0.00E+00
$NH+OH=H_2O+N$	5.00E+11	0.5	2.00E+03
$NH+OH=H_2+NO$	1.60E+12	0.6	1.50E+03
$NH+NH_2=N+NH_3$	1.00E+13	0	2.00E+03
$HO_2+CH_3<=>O_2+CH_4$	1.00E+12	0	0.00E+00
$CH_2+CH_4<=>2CH_3$	2.46E+06	2	8.27E+03

Table 2. AP-HTPB combustion mechanism[a]

$$k = A\ T^b \exp(-E/RT).\ \text{Units: } A\ \text{(mol-cm-s-K)},\ E\ \text{(J/mol)}.$$

M: any metal surface or metallic additive used only as support or catalyst
[a] Kinetic data composed of [8]

Similarly, the combustion process of ammonium perchlorate formulated with hydroxyl terminated polybutadiene was simulated in a perfect stirred reactor (with 70/30 proportion), with variations in the internal chamber pressure (Figure 3 below).

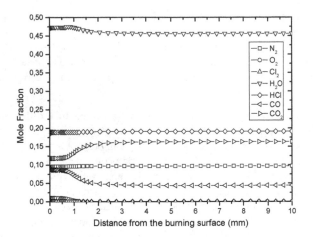

Figure 3. AP/HTPB combustion [7]

The combustion process of AP/HTPB has presented invariable with pressure. This behavior should be attributed to the homogeneous dispersion of AP admist the binder, in the solid phase, and to the lack of this species in relation to the binder (generating lower concentrations of O_2 than necessary). Also, in the gas phases, it is assumed that all of the liquid AP and HTPB present on the condensed phase decompose to form gaseous species; evaporation is not included.

In this simulation, the oxygen molar fraction suffers a decrease (and cancels), according to the reactions with HTPB decomposition products, for the formation of carbon monoxide and dioxide. Also, it is interesting to highlight that in this case the carbon monoxide molar fraction suffers a decrease, because the restriction of oxidizer species makes that the oxygen presented in CO to be also used as oxidizing source, viewing the reactive behavior of this specie. In this simulation, the molar fractions of CO and CO_2 are not null initially, because given the system temperature, HTPB suffers an initial decomposition that should not be discarded, generating both carbon oxides.

There is always the premise in all simulations and all studies that the materials are in a perfect state, flawless. Unfortunately, this is not the reality in most industries or laboratories, when there's low turnover. Therefore, the materials may suffer many different changes in their structure or properties. The main one is the aging process.

The aging process is one of the most significant factors responsible for changes in the activation energy of solid propellants (usually reduction). This phenomenon can be defined as the growth of cross bonds in the polyurethane chain, altering the mechanical properties of traction resistance and elongation, in comparison of the properties just after the fabrication [9]. This aging process can be responsible for the appearance of failures and cracks in the grains, which compromise the propellant performance.

The AP/HTPB composite decomposition and the combustion mechanism have been extensively investigated in the last decades and the appearance of advanced methods of diagnostics, like flash pyrolysis, thermogravimetry and differential scanning calorimetry, led to the ressurgence of the interest. These methods are widely used for the investigation of thermal decomposition of organic materials [10], polymers [11,12], composites [13] and explosives [14].

Kissinger [15] and Ozawa [16] and Flynn [17] demonstrated that differential scanning calorimetry (DSC) technique, based on the linear relation between peak temperature and heating rate, can be used to determine the kinetics parameters of a thermal decomposition (activation energy, rate constant). The Ozawa method is one of the most popular methods for estimating activation energies by linear heating rate and it is the so-called isoconversional method. Thermal analysis cannot be used to elucidate the complete mechanism of a thermal degradation but the dynamic analysis has been frequently used to study the overall thermal degradation kinetics of polymers and composites because it gives reliable information on the frequency factor(A), the activation energy (E) and the overall reaction order [18].

In the present work, the differential scanning calorimetry (DSC) technique and the Ozawa dynamic method were used to determine the kinetic parameters of the aged and non-aged solid propellant, AP/HTPB, thermal decomposition. The Kissinger method for obtaining the activation energy value was also employed for a comparison purpose.

2. Experimental

2.1. Materials and apparatus

AP was obtained from Avibras Indústria Aeroespacial S.A.; HTPB from Petroflex Industry S.A., a subsidiary of Petrobras – Petróleo do Brasil S.A.; IPDI from Merck; DOA from Elekeiroz S.A.. The composite propellant was produced in a batch process of 5 kg mass (pilot plant) using a planetary mixer under vacuum atmosphere during 2 hours. All raw materials are incorporated in HTPB polyol, starting with AP that was classified to a medium size of 300 micrometers. When all ingredients are added to the HTPB polyol, the IPDI curing agent can be mixed to the liquid propellant. The propellant curing process was conducted in a temperature of 60 Celsius during a 120 hs period time.

The synthetic aging process was conducted by exposing the cured propellant formulation to a temperature of 338 K for 300 days in a muffle (FNT-F3-T 6600W) that was monitored day by day during this period.

The polyurethane network was obtained by curing HTPB polymer samples with IPDI (isophore diisocianate) at an [NCO]/[OH] equivalent ratio of 0.95, at 338 K for 120 h. The NCO/OH ratio is defined as the equivalent ratio between the materials containing NCO (IPDI) groups and those containing OH groups (HTPB) and it affects the mechanical properties of cured composite propellant [13,14]. The chemical composition of the propellant was (weight) binder 22% and others 78%. The synthetic aging process was conducted by exposing the propellant formulation to a temperature of 338 K for 300 days.

DSC curves were obtained on a model DSC50 Shimadzu in the temperature range of 298-773 K, under dynamic nitrogen atmosphere (ca. 50 mL/min). Sample masses were about 1.5 mg, and each sample was heated in hermetically sealed aluminum pans. Seven different heating rates were used for the non-aged samples: 10.0, 15.0, 20.0, 30.0, 35.0, 40.0 and 45.0 K min^{-1}; for the aged samples, three different heat rates were used: 30, 35 and 40 K min^{-1}. DSC system was calibrated with indium (m.p.= 429.6 K; ΔH_{fus}=28.54 Jg^{-1}) and zinc (m.p.= 692.6 K).

2.2. Kinetic approach

The method used in the analysis of composite samples was based on DSC experiments in which the temperatures of the extrapolated onset of the thermal decomposition process and the temperatures of maximum heat flow were determined from the resulting measured curves for exothermic reactions. DSC curves at different heating rates, β, for non-aged and aged composite samples are shown in Figs. 4 and 5, respectively.

In order to determine the kinetic parameters of the degradation step Ozawa and Kissinger's methods were applied. They were both derived from the basic kinetic equations for heterogeneous chemical reactions and therefore have a wide application, as it is not necessary to know the reaction order [19] or the conversional function to determine the kinetic parameters. The activation energy determined by applying these methods is the sum of activation energies of chemical reactions and physical processes in thermal decomposition and therefore it is called apparent.

The temperatures of exothermic peaks, T_P, can be used to calculate the kinetic parameters by the Ozawa method [16,17]. These parameters are the activation energy, Ea, and the pre-exponential factor, A, relatives to the decomposition process.

A linear relationship between the heating rate (log β) and the reciprocal of the absolute temperature, T_P^{-1}, may be found and the following linear equation can be established:

$$\log \beta = a.T_P^{-1} + b \qquad (1)$$

where a and b are the parameters of the linear equation: a is -0.4567E/R (slope) and b is a constant (linear coefficient). R is the gas constant.

Assuming that the rate constant follows the Arrhenius law and that the exothermic reaction can be considered as a single step process, the conversion at the maximum conversion rate is invariant with the heating rate when this is linear. Having in account such assumptions, eq.

(1) may be applied to the exothermic peak maximum temperature considering different heating rates [15,19]. Thus carrying out several experiments at different heating rates a plot of log ß vs $1/T_P$ may be done and the activation energy can be estimated directly from the slope of the curve using the following equation derived [14] from the eq.(1):

$$Ea = -2.19 R [d \log ß / d T_P^{-1}]$$ (2)

where $- d \log ß / d T_P^{-1} =$ parameter a (eq.1).

With the same above assumptions, the Kissinger method[8] may be used to calculate the activation energy and the pre-exponential factor from the maximum rate condition which will occur at the maximum exothermic peak temperature, T_P.

The Kissinger method is based on the plot of $\ln (\beta/T_P^2)$ vs. $1/T_P$. Activation energy is calculated from the slope of the curve using the following equation:

$$Ea = R d[\ln \beta/T_P^2] / [d (1/T_P)]$$ (3)

Once time E is known the values of pre-exponential factor, A, are calculated with the equation:

$$A = (\beta E \exp Ea/RT_P) / RT_P^2$$ (4)

The temperature dependence of the specific rate constant k is described by the Arrhenius equation:

$$k = A \exp (-Ea/RT_P)$$ (5)

The kinetic Shimadzu software, based on the Ozawa method, feed with the exothermic peak temperatures and the heating rate data, gives the Arrhenius kinetic parameters (Ea, A) relative to the thermal decomposition of composite and, consequently, with the eq. (5) the overall rate constant can be calculated.

There's also a possibility of using Flynn and Wall methodology with TGA analysis (constant heating rate TGA) [17], once it requires less experimental time, although this method is limited to single-step decompositions and first order kinetics. The approaches are the following:

This first approach requires at least three determinations at different heating rates (Fig. 4 below), following the Arrhenius equation:

$\left(\dfrac{d\alpha}{dt}\right) = A.exp\left(\dfrac{-E\alpha}{RT}\right)(1-\alpha)^n$, where α represents the fraction of decomposition and n is the reaction order.

Re-arranging, the equation turns to:

$E\alpha = \left(\dfrac{-R}{c}\right)\dfrac{d\ln\beta}{d\left(\dfrac{1}{T}\right)}$, where c is a constant for n=1.

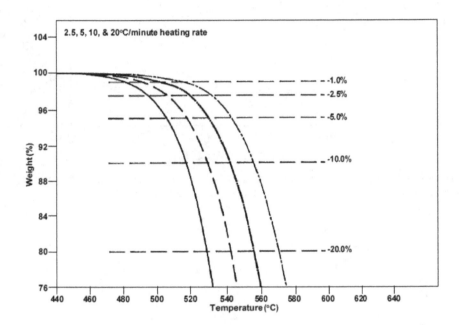

Figure 4. Constant heating rate TGA plots – Flynn & Wall method [20]

From this curve is possible to construct a ln β vs 1/T plot. The slope of this new curve is used
to calculate the activation energy.

3. Results and discussion

The activation energy and kinetic parameters of thermal decomposition of propellant
samples were calculated by Ozawa method using DSC curves at different heating rates: 10.0,
15.0, 20.0, 30.0, 35.0, 40.0 and 45.0 K min⁻¹ for the non-aged ones and 30.0, 35.0 and 40.0 K
min⁻¹ for the aged ones.

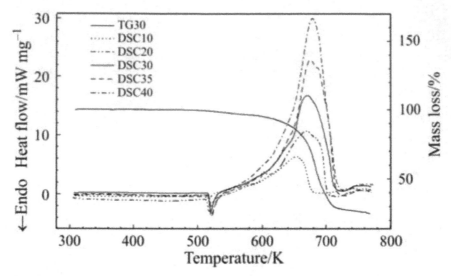

Figure 5. DSC curves of thermal decomposition of non-aged composite samples, AP/HTPB, at the heating rates: 10, 20, 30, 35 and 40 K min^{-1} and TG curve with a heating rate of 30 K min^{-1}

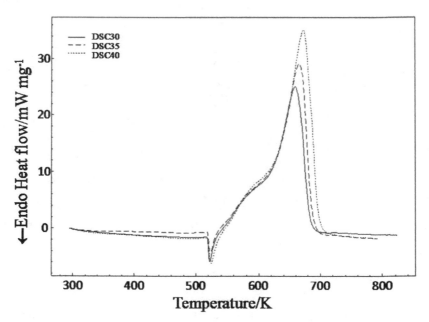

Figure 6. DSC curves of thermal decomposition of aged composite samples, AP/HTPB, at the heating rates: 30, 35 and 40 K min^{-1}

The exothermic events have different maximum temperatures in both cases; the higher the heating rate, higher is the maximum temperature of the peak.

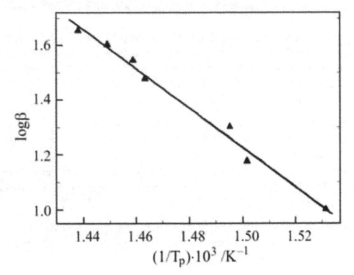

Figure 7. Ozawa plot for AP/HTPB samples at 10, 15, 20, 30, 35, 40 and 45 K min[-1]

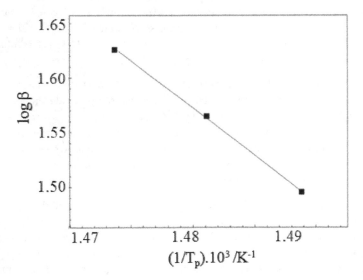

Figure 8. Ozawa plot for aged AP/HTPB samples at 30, 35 and 40 K min[-1]

The DSC curves are presented in Figs. 5 and 6. The DSC curves show that the first stage is endothermic and the second stage is exothermic. The endothermic event is quite similar for the different heating rates used and it shows the same peak temperature. This event occurs around 520 K and it was not considered because it represents a phase transition of ammonium perchlorate (AP) from the orthorhombic to the cubic form [21,22]. Together with DSC curves obtained for different heating rates (non-aged samples), Fig. 1, the TG curve for 30.0 K min^{-1} was included to show that in the region corresponding to the endothermic peak (DSC curves) there is no any weight loss or, at least, it is imperceptible and, the same behavior was observed in all of other TG curves for different heating rates.

Figures 7 and 8 show the plot of log β vs the reciprocal of the absolute temperature relative to each maximum of the exothermic stage. The values of the activation energy were found to be 134.5 kJ mol^{-1} (non-aged samples) and 79.0 kJ mol^{-1} (aged samples). Sell et al.[23] using thermogravimetry at heating rates between 0.5 and 10 K min^{-1} studied the decomposition kinetics of the AP/HTPB propellant samples with isoconversional method and the calculated activation energies are between 100 and 230 kJ mol^{-1}.

From the slope of Kissinger plot (ln (β/T_P^2) vs. 1/T_P) and eq. (3) the activation energy was also calculated and is 126.2 kJ mol^{-1} for the non-aged samples, therefore quite similar to that obtained using the Ozawa method.

The thermal decomposition of solid composite propellant is a multistep process and the reaction mechanism changes with the temperature and, consequently, the activation energy varies with the extent of the reaction. DSC data are used to estimate the activation energies of thermal decomposition of propellant samples because the global decomposition reaction is taken in account. Implicit in any discussion about the decomposition is the fact that the overall process is complex, and any derived rate parameters do not correspond to an elementary single step. TG/DTG results are in agreement with this assumption.

The pre-exponential factor was found to be 2.04 10^{10} min^{-1} (non-aged samples) and 1.29.10^6 min^{-1} (aged samples) and the reaction orders for the global composite decomposition were estimated in 0.7 (non-aged) and 0.6 (aged) by the kinetic Shimadzu software based in the Ozawa method. This value is quite different from the Arrhenius assumption where the reaction order is always considered as 1.0. For practical purposes the Arrhenius parameters, like the corrected reaction order, can be used to estimate the overall rate constant (k) for thermal decomposition using the eq. (5).

The differences found in the kinetics parameters between the original and the aged samples, specially the activation energy (Ea), confirm the practical observation that energetic materials like the composites used in solid propellant rocket motors require less energy to start the combustion process as they age. Besides, considering the heating rate of 40 K min^{-1} for the original and for the aged samples, a reduction in the enthalpy of the decomposition's exothermic phase was observed (2.56 to 1.15 J g^{-1}).

Cohen [24] studied the kinetics of the surface pyrolysis of HTPB and, assuming zero-order kinetics, they found the activation energy of 71 kJ mol^{-1}. Comparison between the activation energies for the propellant decomposition and the activation energies for decomposition of individual ammonium perchlorate (AP) or/and HTPB binder suggests that the overall kinetics of the mass loss is determined by the reaction between the binder and the decomposition products of AP [24].

Ammonium perchlorate is widely used as an oxidizer in energetic composites and it is one of the most important raw materials in propellant formulations where it represents at least 80 % of total mass of composite solid propellants, so its contribution on the thermal decomposition behavior of propellant samples is always very important. The addition of burning rates catalysts like Fe_2O_3 on the propellant formulation alters the thermal decomposition behavior of AP, and consequently the thermal decomposition behavior of the propellant. Shin-Ming [22] showed that the presence of these catalysts compounds reduce the maximum decomposition reaction temperature in AP samples.

Another important aspect of DSC curves is the correlation of maximum temperature of exothermic peak obtained for each heating rate applied to the composite sample during the experiments. This correlation can be used to determine the burning rate characteristics of a composite solid propellant with a specific formulation. The burning rate characteristics are an important ballistic parameter of the energetic composite like solid propellant. Xiao-Bin [25] showed that the burning rates of propellants were very closely related to the exothermic peak temperature of ammonium nitrate (AN) that is used as an oxidizer in smokeless propellant formulation.

In the present work, the DSC curves at different heating rates were obtained for original and synthetically aged samples that have the same raw materials and with the same manufacture process. These conditions are necessary because differences in the raw materials, as ammonium perchlorate (AP) particle size, can affect the thermal decomposition behavior of the composite. In other words, the decomposition mechanism of AP powder of fine particle size differs that of AP of larger particle size.

4. Conclusions

For energetic materials like composite solid propellant, it is critical to use the minimum sample size and low heating rates to avoid the risks to potential damage of the DSC cell resulting in DSC curves with a lot of interferences caused by the detonation behavior of composite samples. In opposition to this criteria, in this study, high heating rates were used (10.0 to 45.0 K min^{-1}), but to compensate this condition very low sample sizes were used ($\approx$ 1.5 mg). Despite these heating rates are not close to the rocket motor chamber conditions (heating rates estimated as 10^6 K s^{-1}) the slower heating rates used in this work allow one to get a better insight into the reaction kinetics mechanisms.

The Ozawa and Kissinger methods demonstrated that differential scanning calorimetry technique, based on the linear relation between peak temperature and heating rate, can be used to determine the kinetics parameters of thermal decomposition reaction of energetic materials giving reproducible results.

The DSC curves do not show any interference and the kinetic data obtained using the maximum temperatures (reciprocal, in K^{-1}) and the respective heating rates are very close to the results found in the literature, at very lower heating rates [26-29].

Author details

R. F. B. Gonçalves, J. A. F. F. Rocco and K. Iha

Instituto Tecnológico de Aeronáutica, CTA, São José dos Campos, S.P., Brasil

Acknowledgement

The authors gratefully acknowledge financial support from FAPESP (Fundação de Amparo à Pesquisa do Estado de São Paulo) and CNPq (Conselho Nacional de Desenvolvimento Científico e Tecnológico) for the fundings and investments.

5. References

[1] Brill T B, Budenz B T (2000) Flash Pyrolysis of Ammonium Perchlorate-Hydroxyl-Terminated-Polybutadiene Mixtures Including Selected Additives, Progress in Astronautics and Aeronautics, AIAA, Vol.185, pp. 3.

[2] Beckstead M W, Puduppakkam K, Thakre P, Yang V (2007) Progress in Energy and Combustion Science, 33, 497.

[3] Boggs T L (1970) AIAA Journal, 8, 5, 867.

[4] Beckstead M W, Puduppakkam K V (2004) Modeling and Simulation of Combustion of Solid Propellant Ingredients Using Detailed Chemical Kinetics, 40th AIAA/ASME/SAE/ASEE Joint Propulsion Conference and Exhibit.

[5] Kubota N (2002) Propellants and Explosives, Wiley-VCH.

[6] Gross M L (2007) Two-dimensional modeling of AP/HTPB utilizing a vorticity formulation and one-dimensional modeling of AP and ADN.

[7] Gonçalves R F B, Rocco J A F F, Machado F B C, Iha K (2012) Ammonium perchlorate and ammonium perchlorate-hydroxyl terminated polybutadiene simulated combustion, Journal of Aerospace Technology and Management, v 4, n1.

[8] Korobeinichev O, Ermolin N, Chernov A, Emel'yanov I (1990) AIAA/SAE/ASME/ASEE 26th Joint Propulsion Conference.

[9] Celina M, Minier L, Assink, R (2002) Development and application tool characterize the oxidative degradation of AP/HTPB/Al propellants in a propellant reability study, Thermochimica Acta, Vol. 384, pp. 343, 349.

[10] Mathot V B F (2001) New routes for thermal analysis and calorimetry as applied to polymeric systems, J. Therm. Anal. Cal., Vol. 64, pp. 15, 35.

[11] Maijling J, Simon P, Khunová V (2002) Optical Transmittance Thermal Analysis of the Poly(Ethylene Terephtalate) Foils, J. Therm. Anal. Cal., Vol. 67, pp. 201, 206.

[12] de Klerk W P C, Schrader, M A, van der Steen A C (1999) Compatibility Testing of Energetic Materials, Which Technique?, J. Therm. Anal. Cal., Vol. 56, pp. 1123, 1131.

[13] Stankovic M, Kapor V, Petrovic S (1999) The Thermal Decomposition of Triple-Base Propellants, J. Therm. Anal. Cal., Vol. 56, pp. 1383-1388.

[14] Jones D E G, Feng H T, Augsten R A, Fouchard R C (1999) Thermal Analysis Studies on Isopropylnitrate, J. Therm. Anal. Cal., Vol. 55, pp, 9, 19.

[15] Kissinger H E (1957) Reaction Kinetics in Differential Thermal Analysis, Anal. Chem., Vol. 29, pp. 1702.

[16] Ozawa T, Isozaki H, Negishi A (1970) A new type of quantitative differential analysis, Thermochimica Acta., Vol. 1, No. 6, pp. 545, 553.

[17] Flynn J H (1966) A quick, direct method for the determination of activation energy from thermogravimetric data, Thermochim. Acta, Vol. 4, pp. 323.

[18] Park J W, Lee H P, Kim H T, Yoo K O (2000) A kinetic analysis of thermal degradation of polymers using a dynamic method, Polym. Degrad.Stabil., Vol. 67, pp. 535.

[19] Ozawa T (2001) Temperature control modes in thermal analysis, J. Therm. Anal. Cal., Vol. 64, 2001, pp. 109, 126.

[20] Sauerbrunn S, Gill P, Decomposition Kinetics using TGA, TA Instruments.

[21] Na-Lu L, Tsao-Fa Y (1991) The thermal behavior of porous residual ammonium percholate, Thermochim. Acta, Vol. 186, pp. 53.

[22] Shin-Ming S, Sun-I C, Bor-Horng W (1993) The thermal decomposition of ammonium perchlorate (AP) containing a burning-rate modifier, Thermochim. Acta, Vol. 223, pp. 135.

[23] Sell T, Vyazovkin S, Wight C A (1999) Thermal decomposition kinetics of PBAN-Binder and composite solid rocket propellants, Combust. Flame, Vol. 119, pp. 174.

[24] Cohen N S, Fleming R W, Derr R L (1974) Role of binders in solid propellant combustion, AIAA Journal, Vol. 6, pp. 212.

[25] Xiao-Bin Z, Lin-Fa H, Xiao-Ping Z (2000) Thermal decomposition and combustion of GAP/NA/Nitrate Ester propellants, Progress in Astronautics and Aeronautics, AIAA, Vol. 185, pp. 413.

[26] Du T (1989) Thermal decomposition studies of solid propellant binder HTPB, Thermochim. Acta, Vol. 138, pp. 189.

[27] Rocco J A F F, Lima J E S, Frutuoso A G, Iha K, Ionashiro M, Matos J R, Suárez-Iha M E V (2004) Thermal degradation of a composite solid propellant examined by DSC – Kinetic study, J. Therm. Anal. Cal., Vol. 75, pp. 551, 557.

[28] Andrade J, Frutuoso A G, Iha K, Rocco J A F F, Bezerra E M, Matos J R, Suárez-Iha M E V (2008) Estudo da decomposição térmica de propelente sólido compósito de baixa emissão de fumaça, Quim. Nova, Vol. 31, No. 2, pp. 301-305.

[29] Andrade J, Iha K, Rocco J A F F, Bezerra E M (2007) Análise térmica aplicada ao estudo de materiais energéticos, Quim. Nova, Vol. 30, No. 4, pp. 952, 956.

Chapter 7

Thermal Analysis of Sulfur and Selenium Compounds with Multiple Applications, Including Anticancer Drugs

Daniel Plano, Juan Antonio Palop and Carmen Sanmartín

Additional information is available at the end of the chapter

1. Introduction

Thermal methodologies are analytical and quantitative methods capable of providing reliable, fast and reproducible results. Thermogravimetry (TG), Differential Scanning Calorimetry (DSC) and Isothermal Titration Calorimetry (ITC) techniques are the chosen methods for several physicochemical determinations.

ITC is the most quantitative means available for measuring the thermodynamic properties of a protein-protein interaction. So, ITC is the calorimetric approach most used to investigate biomolecular interactions. ITC measures the binding equilibrium directly by determining the heat evolved on association of a ligand with its binding partner. In a single experiment, the values of the binding constant (K_a), the stoichiometry (n) and the enthalpy of binding (ΔH_b) are determined. The free energy and entropy of binding are determined from the association constant. The temperature dependence of the ΔH_b parameter, measured by performing the titration at varying temperatures, describes the ΔC_p term. Furthermore, binding of proteins and small molecules to nucleic acids is of course critical to all organisms, playing a role in replication, transcription, translation and DNA repair processes to name just a few. Protein association with nucleic acids has therefore been the subject of much study throughout the years, and ITC has been one of the most common tools used for such investigations. When used in conjunction with complementary techniques such as X-ray crystallography, ITC can provide an informative thermodynamic account of these systems [1]. Besides, ITC is a useful technique in the protein-lipid interactions studies, and two examples in 2008 were the study of the effect of cholesterol on an amphibian antimicrobial peptide interaction with membranes [2], and analysis of the interaction of mammalian bone-marrow derived peptides with model and natural membranes [3]. Finally, ITC is a powerful tool for the pursuit of higher affinity drugs with improved binding specificities [4]. In its simplest form, ITC is a rapid and convenient method for measuring affinities of new leads and optimized compounds. In addition, however, ITC is particularly useful in providing information about the mode of binding. The use of ITC as a general tool in drug design and characterization is exemplified in a study by McKew *et al.* [5] who demonstrated the efficacy of ITC for studying three classes of inhibitor to the cytosolic amphitropic enzyme phospholipase A2 alpha (cPLA2a).

TG is mainly employed to study thermal stability, kinetic parameters and degradation processes for a wide range of materials. DSC allows characterizing protein stability and folding, drug-protein interactions, as well as heat capacity, vapor pressures and polymorphism. Moreover, it has been pointed out the usefulness of DSC technique as a potential tool for the early diagnosis, monitoring and screening of cancer patients [6].

2. Application of thermal analysis to sulfur and selenium compounds with multiple applications

Sulfur (S) and selenium (Se) compounds present several applications in a great variety of fields. We consider the anticancer activity of these compounds as the most important application due to the burden, costs and mortality rates caused by cancer disease. Thus, we will treat in depth the application of thermal techniques to these anticancer compounds in the following section.

Due to the vast applications of S and Se compounds and to the great structural variability in each of these applications, the S and Se compounds are going to be classified according to their structural features.

2.1. Coordinated compounds

The study of the degradation process is one of the most common utility for thermal techniques in the study of metal complex derivatives. Coordination compounds with dithiocarbamates have attracted attention because of their potential biological activity [7-10]. In 2006, a novel dithiocarbamate ligand L (triammonium-*N*-dithiocarboxyiminodiacetate) was synthesized and the thermal decomposition of its cooper (II), niquel (II) and palladium (II) was studied by DSC and thermogravimetry [11]. The authors showed that thermal stability of $(NH_4)_3L$ is low and its decomposition starts with evaporation of an ammonia molecule. Of the three complexes, $Cu(H_2L)_2$ is the least thermally stable. Thermal decomposition of the complexes most likely begins with decarboxylation. It is endothermic up to 500 K, but exothermic oxidation processes are observed above this temperature. Thermal decomposition of the cooper (II) complex is accompanied by its melting and with an exothermic structural rearrangement [11]. During the last two years, these thermal techniques have been used to study the stability and to characterize the degradation process of several dithiocarbamate complexes [12-14]. On the other hand, a study of the degradation process for three novel selenocyanato complexes has been published recently [15]. The authors demonstrated that all compounds decompose in a single heating step without the formation of ligand-deficient intermediates.

Another application of thermal analysis is to study the chemical structure and the structure rearrangements in metallic complexes. The VA main group metal compounds including inorganic and metallorganic complex have showed interesting physical properties, medical and material functions. Some bismuth complexes can be used in medicine, microbiology and pharmacology [16-18]. Two studies with thiourea complexes of antimony and bismuth have evidenced several structure rearrangements or phase transformations for these complexes from 100 to 170 ºC [19, 20].

Several studies have been carried out in order to determine the specific and thermodynamic constants of methionine, which is one of the nine essential amino acids needed by human beings and contains a sulfur atom, and 2-mecaptonicotinic acid complexes [21, 22].

2.2. Glass materials

Chalcogenide semiconductors have been proposed for phase change nonvolatile random access memories, which is becoming the next generation for memory technology [23]. So, the proper description of thermal behavior of semiconducting chalcogenide glasses is crucial to understand their properties and functions. One of the crucial techniques to study the glass transition kinetics is the differential thermal analysis (DTA) and DSC.

Among the chalcogenide systems, selenium and selenium based glassy alloys have been intensively studied due to their wide technical applications, especially in the field of electronics and optoelectronics. A recent study has used the DTA technique to study the glass transition kinetics of the two binary Se-In alloys in comparison with that of pure Se. The glass transition temperature was found to be shifting to a higher value with increasing of heating rates and indium content. It was observed an increase of the stability parameters accompanied with the introduction of In into the Se matrix [24]. Another interesting Se based glassy alloys are Se-Sb alloys owing to their electrical, optical dielectric and thermal properties. Mehta *et al.* have reported the thermal characterization with calorimetric measurements for some Se-Sb alloys [25]. They reported the Hruby number, which is the strong indicator of glass forming tendency, thermal stability parameter and the values of crystallization enthalpy and entropy.

Selenium-tellurium thin films have attractive semiconductors for device application. Se-Te form a continuous series of solid solution and the Se-Te system has an intermediate behavior between pure Se and pure Te. The addition of Te has a catalytic effect on the crystallization of Se. In 2009, the crystallization parameters of the bulk Se-Te chalcogenide glass have been studied using DSC [26]. The values of glass transition temperature, onset crystallization temperature, peak crystallization temperature and enthalpy released with and without laser irradiation for different exposure time have been studied. The films showed indirect allowed interband transition that is influenced by the laser irradiation.

Ternary systems of chalcogenide glasses containing metal elements possess unique optical, electrical and physicochemical properties [27]. The most popular metal is silver and its addition into chalcogenide glasses leads to a drastic change in the physical and chemical properties of the material, for instance, it increases the conductivity by several orders of magnitude and decreases the slope of frequency dependence of alternating current (AC) conductivity. A study conducted by Ogusu *et al.* carried out DSC, X-ray diffraction (XRD) and Raman scattering measurements for $Ag_x(As_{0.4}Se_{0.6})_{100-x}$ glasses with x = 0-35 at.% in order to investigate the crystallization kinetics and the local structure [28]. The DSC curves of the samples with Ag content x = 15-35 at.% were obtained at various heating rates for different Ag contents and two or three exothermic peaks for the crystallization were found depending on the Ag content. Furthermore, the dimension of crystal growth of sample particles and activation energy were determined using Matusita's equation to analyze the DSC data. It was found that the surface and bulk crystallization take place depending on the Ag content and peak crystallization temperatures [28].

Glassy selenium has low sensitivity and thermal instability. These properties can be improved by alloying of some elements into selenium matrix, such as arsenic [29] and antimony [30, 31]. The proper description of thermal behavior of these glasses is important for understanding their properties and applications. Recently, the thermal properties and structure of As_xSe_{100-x} and Sb_xSe_{100-x} glass-forming systems (x = 0, 1, 2, 4, 8 and 16) were reported by conventional and StepScan DSC and Raman spectroscopy [32]. Among these thermal properties, the authors studied the glass transition temperature and the crystallization of undercooled melts. So, the glass transition temperature for As_xSe_{100-x} system increases almost linearly with increasing As content from 40 up to 93 °C, because the glass structure becomes more stable due to cross-linking of Se chains by As. Nevertheless, the glass transition temperature of Sb_xSe_{100-x} changes only slightly from 40 to 48 °C [32]. Concerning to the study of crystallization of undercooled melts, it was found that only selenium crystallizes from undercooled melts of As-Se system and its tendency to crystallize decreases markedly with increasing As content, for arsenic content higher than 4 at.% no crystallization was observed. In the case of Sb-Se system Sb_2Se_3 crystallizes in the first step followed by trigonal selenium crystallization from non-stoichiometric undercooled melt [32].

Another technologically important ternary system of chalcogen elements are the infrared transmitting glasses based on Ge-Sb-Se because they are good transmitters of radiation in the 2-16 µm wavelength region. The applications include fabrication of optical components like IR lenses, windows and filter used in thermal imaging systems. The Sb-Ge-Se films result sensitive for the UV exhibit mechanical, optical and structural changes [33, 34]. An understanding of the glass forming tendency and crystallization kinetics in these chalcogenide materials is very important to develop them for applications based on the amorphous to crystallization phase change and vice versa. So, one report evaluated the glass-forming ability of some alloys in $Sb_xGe_{25-x}Se_{75}$ ($0 \leq x \leq 10$) system by using various thermal stability criteria, based on characteristic temperatures [35]. It was observed that the thermal stability decrease with increasing Sb content in the glassy system.

2.3. Inorganic mixtures

Several reports have been published concerning to the solubility and thermal characterization of various metal-selenite systems. For example, some manganese(II) selenite are used for coloring glasses, enamel and glazes. On the basis of $MnSeO_3$, two manganese selenides (α-MnSe and $MnSe_2$) were obtained having very interesting semiconductor properties [36]. Vlaev et al. have studied the crystallization fields of manganese(II) selenites in the system $MnSeO_3$-SeO_2-H_2O in the temperature interval 25-300 ºC and characterized the observed phases [37]. Previously to this article, the same author reported the crystallization fields and the characterization of the observed phases for the system $NiSeO_3$-SeO_2-H_2O [38]. Another article studied the phase equilibrium in the system CdO-SeO_2-H_2O at 25 and 100ºC and the thermolysis mechanism of the compounds obtained.

The ytterbium selenites can serve as initial substances for obtaining selenides and oxyselenides having valuable photoconductive and superconductive properties. So, Gospodinov et al. have studied the solubility isotherm of the three-component system Yb_2O_3-SeO_2-H_2O at 100 ºC [39]. Furthermore, they have performed simultaneous TG and DTA curves of the compounds obtained in its fields of crystallization and the mechanism of the thermal decomposition [39].

Alkali metal sulfates, selenite and phosphate tellurate compounds having the formula $M_2XO_4Te(OH)_6$, where M is the metal and X is S, Se or P, form a broad families with interesting properties, such as superprotonic conduction and ferroelectricity [40-42]. So, synthesis, calorimetric and conductivity studies of new mixed solution of rubidium sulfate selenate tellurate [43] and thallium selenate tellurate [44] have been carried out.

2.4. Miscellaneous compounds

In the last two years, several thermal and structural investigations on crystal structures with thiourea have been carried out. The thermal decomposition of crystal structures with bisthiourea derivatives has been studied by TG-DSC [45]. Another study reported the growth and characterization of a new non-linear organometallic crystal (potassium thiourea thiocyanide or PTT) [46]. The TG curve showed the complete decomposition of PTT between 176 and 1000 ºC in three steps with corresponding three DTA peaks.

Some selenoesters present promising photophysical properties for optical device applications such as emissive liquid crystal displays (LCDs), polarized organic lasers and anisotropic Light-emitting diodes (LEDs). Rampon et al. have reported the synthesis and the study of the liquid crystalline and fluorescent properties of novel selenoesters [47]. So, these compounds were fluorescent in the blue region and exhibited their stability and liquid crystalline properties over a large range of temperatures. Moreover, these compounds showed a rich phase polymorphism.

Cooper chalcogenides are considered as promising in electronic technology due to their physicochemical properties [48, 49]. Chrissofis et al. [50] have reported the thermal behavior of samples with very slight divergence from stoichiometry ($Cu_{2-x}Se$). Also, they have studied the nature of the transformation with non-isothermal measurements at different heating and cooling rates.

Oligothiophenes and polythiophenes are another sulfur compounds that have attracted much attention due to their unusual electric and nonlinear optical properties as interesting materials for organic electronics and optoelectronics, LEDs, field-effect transistors, thin-film transistors... So, the relative stabilities of 2,2′- and 3,3′-bithiophenes (the main building blocks of these conducting organic materials) have been evaluated by experimental thermochemistry [51].

3. Application of thermal analysis to sulfur and selenium compounds with anticancer activity

In the last decade, among the wide range of compounds tested as potential anticancer agents, several structurally diverse derivatives that contain a sulfur or selenium template have been reported and have generated growing interest. For that reason, in this chapter we have focused on some relevant thermal studies in sulfur and selenium compounds with anticancer activity.

3.1. Sulfur amino acids and cysteine cathepsins.

The human family of cysteine cathepsins are a family of lysosomal proteases and has 11 members (cysteine cathepsin B, C, F, H, K, L, O, S, V, W and X), which share a conserved active site that is formed by cysteine, histidine and asparagine residues. Cysteine cathepsins are often upregulated in various human cancers, and have been implicated in distinct tumorigenic processes such as angiogenesis, proliferation, apoptosis and invasion. During cancer progression, cathepsins are often translocated to the cell surface of tumor cells or are secreted into the extracellular milieu, where they can promote tumor invasion through several possible mechanisms. Causal roles for cysteine cathepsins in cancer have been demonstrated by pharmacological and genetic techniques. This includes functional downregulation of cysteine cathepsin activity by increasing expression of endogenous inhibitors and administration of small-molecule cysteine protease inhibitors. Besides, causal roles for specific cysteine cathepsins in cancer have been demonstrated by downregulating their expression or crossing mouse models of cancer with mice in which the cysteine cathepsin has been genetically ablated. These studies have identified roles for cysteine cathepsins in both tumor cells and tumor-associated cells such as endothelial cells and macrophages.

Taking into account the causal roles for cysteine cathepsins, which present a cysteine residue in the active site, in cancer and the fact that the thiol-disulfide interchange reaction is important to a number of subjects in biochemistry, thermodynamic data regarding the relative energetics of the thiol and disulfide functional groups is essential for the understanding of the driving force and mechanism of biochemical processes. Temperature-induced changes in crystalline amino acids are of interest for their properties and because they reveal the intrinsic motions of these structural fragments and their contribution to the dynamic properties of proteins.

So, a thermophysical study of the sulfur containing amino acids L-cysteine and L-cystine by DSC has been reported [52]. Heat capacities of both compounds were measured in the temperature interval from T = 268 K to near their respective melting temperatures.

Furthermore, a solid-solid phase transition close to the melting point is only observed in the L-cysteine. Additionally, several polymorphic forms have been reported for both compounds. L-cysteine crystallizes in the monoclinic and orthorhombic forms and has been structurally characterized [53, 54]. Phase transitions have been detected when lowering the temperatures [55] and also when decreasing pressures up to 4.2 GPa and decreasing to 1-7 GPa [56]. L-cystine crystallizes in the tetragonal and hexagonal forms and has also been studied at ambient [57, 58] and at low temperature [59] and at high pressures although no solid-solid phase transition has been detected.

3.2. Metal complexes of sulfur compounds

During the past decade the study of mixed sulfur donor ligand complexes with main group metals has made a progressive development due to the development of new analytical and structural techniques [60-62]. These complexes present potential applications in areas such as fast ion conductivity, photocatalysis and electro-optics, among others as well as several biochemical applications [63-65].

One of these promising complexes is antimony(III) bis(pyrrolidinedithiocarbamato) alkyldithiocarbonates. The link of two active ligands was the rational design used for the design of these complexes. So, pyrrolidine dithiocarbamates which represent a class of antioxidants mediate a wide variety of effects in biological systems[66]. It is a multipotent synthetic compound well known for its metal chelation property and one of the most potent and specific NF-kB inhibitor [67]. Besides, antimony metal containing compounds are commonly used to treat parasitic infections and exhibit a broad spectrum of chemotherapeutic applications and cytotoxic activities. So, a study has reported the synthesis, spectroscopic, thermal and structural behavior of antimony(III) bis(pyrrolidinedithiocarbamato)alkyldithiocarbonates [12]. Thermogravimetric studies not only allows to determine purity and thermal stability of the complex but also composition of the complex as well which it is observed during different steps of weight losses as a fragment formed in different temperature ranges [12].

Another type of complexes with a potent anticancer activity which were designed using the link of two active ligands, are the palladium (II) and platinum (IV) complexes with active sulfur ligands. The use of antitumor drugs based on platinum(II) metal complexes, cisplatin and its analogues carboplatin and oxaliplatin is limited by two factors: installation of tumor drug resistance and severe adverse effects [68, 69]. Therapeutic strategies are oriented towards the development of new platinum- and non-platinum-based antitumor drugs with higher efficiency, reduced general toxicity and broader spectrum of activity [70]. Sulfur-containing molecules are studied as chemoprotectors in platinum-based chemotherapy. Dithiocarbamates have attracted particular attention for the use of chemical modulation of cisplatin nephrotoxicity [71-73].

The thermal behaviour of Pd(II) complexes with some dithiocarbamate derivatives was studied in order to establish the coordination mode of the ligand, to test the thermal stability [11] or to understand the effect of the alkyl chain attached to the nitrogen atom over the thermochemical parameters of the complexes [74, 75]. A series of Pd(II) complexes with *tert*-butylsarcosinedithiocarbamate [76], ethylsarcosinedithiocarbamate and 2-/3-picoline [77], dithiocarbamates and various amines [78] were developed as antitumor agents with low nephrotoxicity. Thermogravimetric analysis was used for the characterization of these compounds. Taking into accounts the kinetic inertness, high activity, low toxicity and suitability for oral administration of Pt(IV) complexes, some Pt(IV) complexes with dithiocarbamates have been synthesized. Morpholine dithiocarbamate, aniline dithiocarbamate and N-(methyl, cyclohexyl) dithiocarbamate alone [79] or with triphenyl phosphine as second ligand [80] were used in order to obtain Pt(IV) complexes with antitumour activity.

In 2012, Uivarosi *et al.* have reported the thermal and spectral studies of palladium(II) and platinum(IV) complexes with bis(dimethylthiocarbamoyl)sulphide and bis(diethylthiocarbamoyl)disulphide [81]. TG experiments revealed the nature of complex species as hydrated or anhydrous. Thermal decomposition of coordinated organic ligands occurs in one or two exothermic stages, the final residue being in all cases the free metal (Pd or Pt).

Other complexes with anticancer activity are de ruthenium (III) complexes with sulfur ligands. Ruthenium complexes with dimethyl sulfoxide (dmso) showed selective antitumor properties in preclinical testing [82]. Biological studies in *cis*- and *trans*-RuX$_2$(dmso)$_4$ complexes (X = Cl and Br) refer to different tumor toxicity and anti-metastasis properties of the isomers [83]. Dmso can be coordinated to ruthenium as a metal center either through the sulfur (dmso-S) or through the oxygen atom (dmso-O). Dmso provides a moderate acceptor site for π-electron donors and bound through sulfur stabilizes ruthenium in lower Ru(II) oxidation state, more reactive toward tumor cells [84]. The biological activity of complexes can be modified by addition or change of the ligands. Phenothiazines and their N-alkyl derivatives are themselves biological active compounds, suitable to take part in complex formation. Moreover, they exhibit a strong in vitro antitumor activity in numerous and various tumor cell lines [85].

Recently, thermal decomposition of chlorpromazine hydrochloride (CP·HCl), trifluoperazine dihydrochloride (TF·2HCl) and thioridazine hydrochloride (TR·HCl), and the ruthenium complexes with dimethyl sulfoxide (dmso) of composition [RuCl$_2$(dmso)$_4$] and L[RuCl$_3$(dmso)$_3$]·xEtOH, L = CP·HCl, TF·2HCl or TR·HCl is described [86]. The phenothiazines are stable to temperature range of 200–280 ºC with an increasing stability order of TF·2HCl < CP·HCl < TR·HCl. The decomposition of all the compounds takes place in superposing steps.

3.3. Alkylimidothio- and alkylimidoselenocarbamates

During the last five years, our research group reported the promising and potent anticancer effects for several alkylimidothio- and alkylimidoselenocarbamate derivatives [87-89]. These compounds showed a remarkable cytotoxic activity *in vitro* against prostate cancer cells and other several cancer cell lines. One of these derivatives, the quinoline imidoselenocarbamate

EI201, inhibits the PI3K/AKT/mTOR pathway, which is persistently activated and contributes to malignant progression in various cancers, and contributes to the loss of maintenance of the selfrenewal and tumorigenic capacity of cancer stem cells. This compound (EI201) suppressed almost 80% prostate tumor growth *in vivo* ($p < 0.01$) compared to controls at a relatively low dose (10 mg/kg) in a mouse xenograft model [90].

Degradation and fusion temperatures for 20 of these anticancer derivatives were determined using TG and DSC [91]. Analysis of the thermal data indicated that: (a) in general, sulfur compounds are more stable than selenium compounds; (b) the pyridine ring diminished stability of sulfur and selenium compounds much more than the carbocyclic aromatic rings did; (c) selenomethyl derivatives are more stable than selenoethyl and selenoisopropyl compounds; (d) a chlorine atom on selenocompounds has surprising effects. So, the presence of intermolecular bonds was pointed out between chlorine atom and selenium atom [91]. With regard some substituents present on aromating ring and the ramification and length of chain, it can be concluded that the presence of electron-withdrawing groups in selenocompound structures improves their stability. Besides, selenomethyl derivatives are more stable than selenoethyl and selenoisopropyl compounds [91].

The determination of the polymorphism of a substance is of great importance due to the strong influence of the crystalline form on the physicochemical properties, bioavailability and stability of drugs [92], and, in some compounds with biological activity, can even become metastable forms, being twice as active as the stable form [93]. So, our research group has carried out the study of the physicochemical properties of polymorphic forms of a serie of alkylimidothio- and alkylimidoselenocarbamate derivatives with a combination of DSC, thermomicroscopy and X-ray diffractometry [94]. In this study we observed that polymorphs could be formed when the compounds are heated above their melting points. The results showed that there are four types of thermal behavior for alkylimidothio- and alkylmidoselenocarbamate derivatives: (a) compounds which do not evidence any polymorphic forms (behavior I); (b) compounds which solidify into an amorphous solid form (behavior II); (c) compounds which present a new polymorphic form at a Tonset lower than the original one (behavior III); (d) finally, compounds which have three polymorphic forms with three different Tonset values (behavior IV). Calorimetric studies demonstrated that sulfur and selenium analogs have the same thermal behavior. So, the different thermal behaviors observed for these alkylimidothio- and alkylimidoselenocarbamates are caused by the substituent groups in the aromatic ring, although there is no relationship between electron-withdrawing and electron-donating groups and the thermal behavior.

3.4. Case study: Thermal analysis of selenyl acetic derivatives.

In 2009, our research group reported the cytotoxic and antiproliferative activities *in vitro* of selenyl acetic derivatives against several cancer cell lines [95]. Considering the structure of these derivatives and their inefficacy to induce apoptosis and to affect to cell cycle, we decided to perform a thermal analysis for these derivatives. So, we carried out a thermal stability and calorimetric studies for some of these anticancer selenyl acetic acids (**Figure 1**).

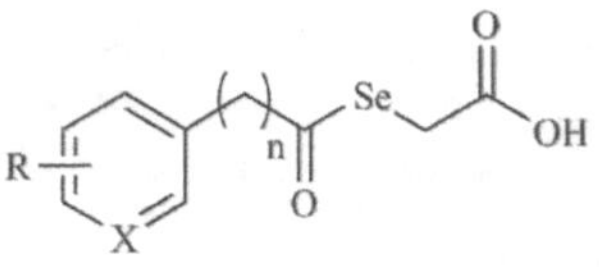

Compound 1: R = 2-Cl;	X = N; n = 0
Compound 2: R = H;	X = C; n = 0
Compound 3: R = 3,5-diOCH$_3$;	X = C; n = 0
Compound 4: R = 4-CF$_3$;	X = C; n = 0
Compound 5: R = 4-CN	X = C; n = 0
Compound 6: R = 4-CH$_3$;	X = C; n = 0
Compound 7: R = H;	X = C; n = 1
Compound 8: R = Phenyl;	X = C; n = 0

Figure 1. General structure for studied selenyl acetic acids.

3.4.1. Thermal stability studies

The thermogravimetric studies were carried out with a Perkin-Elmer TGA-7. The thermobalance was calibrated with alumel and nickel at 10 ºC min⁻¹. The calibration of the oven temperatures was carried out automatically. Mass calibration was carried out with a certified mass of 10 mg (ASTM E617).

The calorimeter was calibrated with indium and zinc (provided by Perkin-Elmer and fabricated according to guideline ISO35) at 10 ºC min⁻¹ and a nitrogen flow of 20 mL min⁻¹. The gases connected to the equipment were nitrogen and air with a purity of 99.999%.

Thermogravimetric analyses were carried out under nitrogen atmosphere with a gas flow of 40 mL min⁻¹ at 10 ºC min⁻¹, using a sample of approximately 3 mg.

All the compounds sublimated before the degradation process start. So, it is not possible to study the thermal stability of these compounds using thermogravimetric techniques.

3.4.2. Calorimetric studies

The calorimetric studies were carried out with a Perkin-Elmer DSC Diamond. Calorimetric analyses were carried out in aluminium capsules for volatiles of 10 μL, at a heating rate of 10 ºC min⁻¹, using a sample of approximately 3 mg, in order to establish the T_{onset} and the enthalpy of fusion ΔH_f. All of the experiments were performed at least three times and the values were expressed as mean ± standard deviation.

The obtained data (**Table 1**) allow us to point out the following calorimetric behaviors:

1. Regarding to the T_{onset} values:
a. The substitution on the aromatic ring causes an increase in the fusion temperatures of these compounds.
b. The inclusion of a methylene group between aromatic ring and carbonyl group seem to lead to a diminution in the T_{onset} values.
c. The presence of groups such as cyano and chloro on the ring, which can form hydrogen bonds, significantly increase the fusion temperatures of these derivatives.
2. Regarding to the enthalpy of fusion values:

a. Compounds with phenyl ring (2 and 7) possess values for enthalpy of fusion higher than T_{onset} ones. It could be caused by the presence of strong π-π stacking interactions between two molecules in the crystal packaging.

b. Compounds with groups that can establish hydrogen interactions, such as CN and Cl, substituted over the ring present the highest enthalpy values, owing to these hydrogen interactions.

c. The substitution on the *para* position of the phenyl ring with groups that cannot establish hydrogen interactions significantly diminished the enthalpy values. It seems that these substituents alter the electronic distribution over the ring, affecting to the π-π stacking interactions strength.

Reference	T_{onset} (°C ± SD)	ΔH_f (Jg^{-1})
Compound 1	157.2 ± 0.2	126.8 ± 4.8
Compound 2	83.3 ± 0.6	92.3 ± 2.2
Compound 3	117.4 ± 0.3	108.3 ± 5.5
Compound 4	107.7 ± 1.1	73.7 ± 1.0
Compound 5	146.5 ± 0.1	112.3 ± 1.7
Compound 6	92.2 ± 0.2	78.7 ± 0.8
Compound 7	73.8 ± 0.3	74.9 ± 7.2
Compound 8	130.1 ± 0.4	105.9 ± 4.6

Table 1. T_{onset} and enthalpy of fusion values for selenyl acetic acid derivatives studied.

The calorimetric data (**Table 1**) demonstrated that compounds 1 and 5 present a very significant higher T_{onset} value for the fusion process than the other selenyl acetic derivatives. Both compounds present two groups substituted in the *para* position of the aromatic ring that can act as hydrogen bonding donors. So, these groups could form a different hydrogen bond interaction with the proton of carboxylic acid group (**Figure 2**) and this interaction should be stronger than the interactions established in the rest of the compounds.

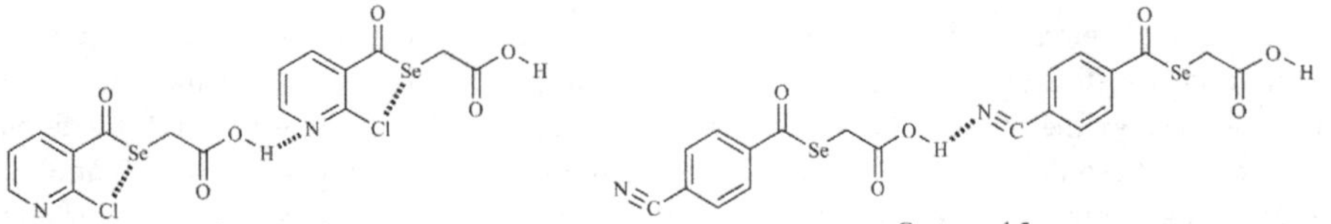

Compound 1   Compound 5

Figure 2. Possible interactions in the crystal packaging for compounds 1 and 5.

If we compare the calorimetric data for selenyl acetic derivatives with other organoselenium compounds (alkylimidoselenocarbamates) synthesized and published by our research group [91], we observed that the T_{onset} values for the first ones were significantly lower that their imidoselenocarbamate analogs (**Table 2**). The selenyl acetic derivatives are smaller molecules than imidoselenocarbamates and hence they possess lower T_{onset} values. Nevertheless, the enthalpy of fusion values for selenyl acetic acids are significantly higher compared with the imidoselenocarbamates in all cases (**Table 2**). These data seem to point to the existence of intramolecular hydrogen bonds for selenyl acetic derivatives (**Figure 3**).

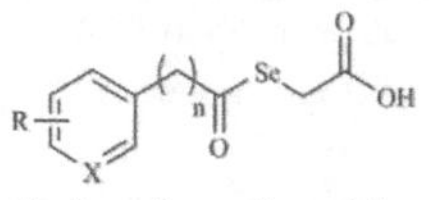

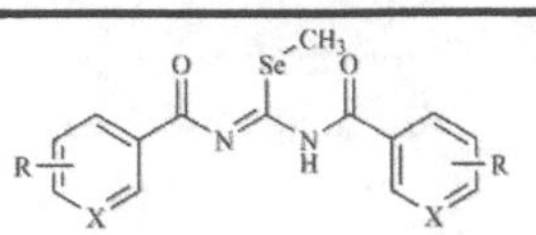

Ref.	R	X	n	Selenyl acetic acids		Imidoselenocarbamates	
				T_{onset} (°C ± SD)	ΔH_f (Jg⁻¹)	T_{onset} (°C ± SD)	ΔH_f (Jg⁻¹)
1	2-Cl	N	0	157.2 ± 0.2	126.8 ± 4.8	186.8 ± 0.2	42.4 ± 1.3
2	H	C	0	83.3 ± 0.6	92.3 ± 2.2	139.5 ± 0.2	30.6 ± 0.6
3	3,5-diOCH₃	C	0	117.4 ± 0.3	108.3 ± 5.5	163.6 ± 0.3	39.6 ± 0.3
4	4-CF₃	C	0	107.7 ± 1.1	73.7 ± 1.0	172.9 ± 0.3	30.7 ± 0.8
5	4-CN	C	0	146.5 ± 0.1	112.3 ± 1.7	219.5 ± 2.8	44.4 ± 14.3
6	4-CH₃	C	0	92.2 ± 0.2	78.7 ± 0.8	148.5 ± 0.3	32.7 ± 4.9
7	H	C	1	73.8 ± 0.3	74.9 ± 7.2	---	---
8	Phenyl	C	0	130.1 ± 0.4	105.9 ± 4.6	---	---

Table 2. T_{onset} and enthalpy of fusion values for selenyl acetic acids and imidoselenocarbamates.

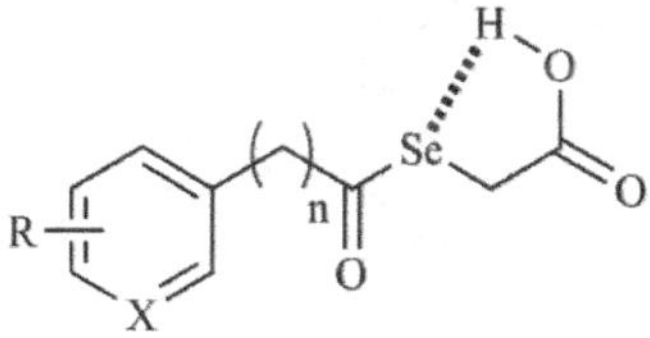

Figure 3. Intramolecular hydrogen bonds for selenyl acetic derivatives.

4. Conclusions

Sulfur (S) and selenium (Se) compounds present several applications in a great variety of fields and the study of their thermal behavior is important for their usefulness in these applications. So, the thermal data are necessary to understand the properties and functions for most of these derivatives. The application of the thermal techniques to these S and Se compounds allows, among others: (a) the study of degradation process, as well as the quantification of the purity and composition for coordination compounds; (b) the characterization of thermal behavior for semiconducting glasses; (c) the determination of solubility isotherms and the field of crystallization of inorganic metal-selenite mixtures.

In the last decade, among the wide range of compounds tested as potential anticancer agents, several structurally diverse derivatives that contain a sulfur or selenium template have been reported and have generated growing interest. For that reason, in this chapter we have focused on some relevant thermal studies in sulfur and selenium compounds with anticancer activity. The thermal techniques, and particularly the DSC, are especially useful in these compounds since they can point out future issues in its pharmaceutical development. The determination of the polymorphism of a substance is of great importance due to the strong influence of the crystalline form on the physicochemical properties, bioavailability and stability of drug, and, in some compounds with biological activity, can even become metastable forms, being twice as active as the stable form.

Finally, we report the unpublished thermal analysis data for eight selenyl acetic acid derivatives, which possess cytotoxic activity *in vitro* against several cancer cell lines. All the compounds sublimated before the degradation process start. So, it is not possible to study the thermal stability of these compounds using thermogravimetric techniques. Nevertheless, the obtained results for calorimetric studies allow to point out some calorimetric behaviors concerning to their stability in the fusion process: (a) the substitution on the aromatic ring causes an increase in the fusion temperatures of these compounds; (b) the inclusion of a methylene group between aromatic ring and carbonyl group seem to lead to a diminution in the T_{onset} values; (c) the presence of groups such as cyano and chloro on the ring, which can form hydrogen bonds, significantly increase the fusion temperatures of these derivatives.

Author details

Daniel Plano*
Synthesis Section, Department of Organic and Pharmaceutical Chemistry, University of Navarra, Pamplona, Spain
Department of Pharmacology, Penn State Hershey College of Medicine, Hershey, PA, USA

Juan Antonio Palop and Carmen Sanmartín
Synthesis Section, Department of Organic and Pharmaceutical Chemistry, University of Navarra, Pamplona, Spain

5. References

[1] Zhou Y Z, Larson J D, Bottoms C A, Arturo E C, Henzl M T, Jenkins J L, Nix J C, Becker D F, Tanner J J (2008) Structural Basis of the Transcriptional Regulation of the Proline Utilization Regulon by Multifunctional PutA. Journal of Molecular Biology. 381: 174-188.

[2] Verly R M, Rodrigues M A, Daghastanli K R P, Denadai A M L, Cuccovia I M, Bloch C, Frezard F, Santoro M M, Pilo-Veloso D, Bemquerer M P (2008) Effect of Cholesterol on the Interaction of the Amphibian Antimicrobial Peptide DD K with Liposomes. Peptides. 29: 15-24.

[3] Andrushchenko V V, Aarabi M H, Nguyen L T, Prenner E J, Vogel H J (2008) Thermodynamics of the Interactions of Tryptophan-Rich Cathelicidin Antimicrobial Peptides with Model and Natural Membranes. Biochimica Et Biophysica Acta-Biomembranes. 1778: 1004-1014.

* Corresponding Author

[4]   Chaires J B. In *Annual Review of Biophysics*, 2008; Vol. *37*, pp 135-151.

[5]   McKew J C, Lee K L, Shen M W H, Thakker P, Foley M A, Behnke M L, Hu B, Sum F W, Tam S, Hu Y, Chen L, Kirincich S J, Michalak R, Thomason J, Ipek M, Wu K, Wooder L, Ramarao M K, Murphy E A, Goodwin D G, Albert L, Xu X, Donahue F, Ku M S, Keith J, Nickerson-Nutter C L, Abraham W M, Williams C, Hegen M, Clark J D (2008) Indole Cytosolic Phospholipase a(2) Alpha Inhibitors: Discovery and in Vitro and in Vivo Characterization of 4-{3-5-Chloro-2-(2-{(3,4-dichlorobenzyl)sulfonyl amino}ethyl)-1-(diphenylmethyl)-1H-indol-3-yl propyl}benzoic Acid, Efipladib. Journal of Medicinal Chemistry. 51: 3388-3413.

[6]   Garbett N C, Mekmaysy C S, Helm C W, Jenson A B, Chaires J B (2009) Differential Scanning Calorimetry of Blood Plasma for Clinical Diagnosis and Monitoring. Experimental and Molecular Pathology. 86: 186-191.

[7]   Manoussakis G, Bolos C, Ecateriniadou L, Sarris C (1987) Synthesis, Characterization and Anti-Bacterial Studies of Mixed-Ligand Complexes of Dithiocarbamato−Thiocyanato and Iron(III), Nickel(II), Copper(II) and Zinc(II). European Journal of Medicinal Chemistry. 22: 421-425.

[8]   Montagner D, Marzano C, Gandin V (2011) Synthesis, Characterization and Cytotoxic Activity of Palladium (II) dithiocarbamate Complexes with Alpha,Omega-Diamines. Inorganica Chimica Acta. 376: 574-580.

[9]   Chen Y-W, Chen K-L, Chen C-H, Wu H-C, Su C-C, Wu C-C, Way T-D, Hung D-Z, Yen C-C, Yang Y-T, Lu T-H (2010) Pyrrolidine Dithiocarbamate (PDTC)/Cu Complex Induces Lung Epithelial Cell Apoptosis through Mitochondria and ER-Stress Pathways. Toxicology Letters. 199: 333-340.

[10] Mohamed G G, Ibrahim N A, Attia H A E (2009) Synthesis and Anti-Fungicidal Activity of Some Transition Metal Complexes with Benzimidazole Dithiocarbamate Ligand. Spectrochimica Acta Part A-Molecular and Biomolecular Spectroscopy. 72: 610-615.

[11] Leka Z B, Leovac V M, Lukic S, Sabo T J, Trifunovic S R, Szecsenyi K M (2006) Synthesis and Physico-Chemical Characterization of New Dithiocarabamato Ligand and Its Complexes with Copper(II), Nickel(II) and Palladium(II). Journal of Thermal Analysis and Calorimetry. 83: 687-691.

[12] Chauhan H, Bakshi A (2011) Synthetic, Spectroscopic, Thermal, and Structural Studies of Antimony(III) bis(pyrrolidinedithiocarbamato)alkyldithiocarbonates. Journal of Thermal Analysis and Calorimetry. 105: 937-946.

[13] Onwudiwe D C, Ajibade P A (2011) Synthesis, Characterization and Thermal Studies of Zn(II), Cd(II) and Hg(II) Complexes of N-Methyl-N-Phenyldithiocarbamate: The Single Crystal Structure of [(C6H5)(CH3)NCS2]4Hg2. International Journal of Molecular Sciences. 12: 1964-1978.

[14] Onwudiwe D C, Ajibade P A, Omondi B (2011) Synthesis, Spectral and Thermal Studies of 2,2'-Bipyridyl Adducts of Bis(N-Alkyl-N-Phenyldithiocarbamato)Zinc(II). Journal of Molecular Structure. 987: 58-66.

[15] Wriedt M, Naether C (2011) Synthesis, Crystal Structures, Thermal and Magnetic Properties of New Selenocyanato Coordination Polymers with Pyrazine as Co-Ligand. Zeitschrift Fur Anorganische Und Allgemeine Chemie. 637: 666-671.

[16] Köpf-Maier P, Klapötke T (1988) Antitumor Activity of Some Organomettalic Bismuth(III)Thiolates. Inorganica Chimica Acta. 152: 49-52.

[17] Cantos G, Barbieri C L, Iacomini M, Gorin P A J, Travassos L R (1993) Synthesis of Antimony Complexes of Yeast Mannan and Mannan Derivatives and Their Effect on Leishmania-Infected Macrophages. Biochemical Journal. 289: 155-160.

[18] Briand G G, Burford N (1999) Bismuth Compounds and Preparations with Biological or Medicinal Relevance. Chemical Reviews. 99: 2601-2658.

[19] Luan S R, Zhu Y H, Jia Y Q, Cao Q (2010) Characterization and Thermal Analysis of Thiourea and Bismuth Trichloride Complex. Journal of Thermal Analysis and Calorimetry. 99: 523-530.

[20] Zhong G Q, Luan S R, Wang P, Guo Y C, Chen Y R, Jia Y Q (2006) Synthesis, Characterization and Thermal Decomposition of Thiourea Complexes of Antimony and Bismuth Triiodide. Journal of Thermal Analysis and Calorimetry. 86: 775-781.

[21] Swain T (2012) Synthesis and Thermal Characterization of Sulfur Containing Methionine Bridged Cobalt(III) and Copper(II) Complex. Journal of Thermal Analysis and Calorimetry. 109: 365-372.

[22] Yang Q, Chen S, Gao S (2007) Syntheses and Thermal Properties of Some Complexes with 2-Mercaptonicotinic Acid. Journal of Thermal Analysis and Calorimetry. 90: 881-885.

[23] Lee M L, Shi L P, Tian Y T, Gan C L, Miao X S (2008) Crystallization Behavior of $Sb_{70}Te_{30}$ and $Ag_3In_5Sb_{60}Te_{32}$ Chalcogenide Materials for Optical Media Applications. Physica Status Solidi (a). 205: 340-344.

[24] Kotkata M F, Mansour S A (2011) Study of Glass Transition Kinetics of Selenium Matrix Alloyed with up to 10% Indium. Journal of Thermal Analysis and Calorimetry. 103: 555-561.

[25] Mehta N, Tiwari R S, Kumar A (2006) Glass Forming Ability and Thermal Stability of Some Se–Sb Glassy Alloys. Materials Research Bulletin. 41: 1664-1672.

[26] Bahishti A A, Majeed Khan M A, Patel B S, Al-Hazmi F S, Zulfequar M (2009) Effect of Laser Irradiation on Thermal and Optical Properties of Selenium–Tellurium Alloy. Journal of Non-Crystalline Solids. 355: 2314-2317.

[27] Ohto M, Itoh M, Tanaka K (1995) Optical and Electrical Properties of Ag-As-S Glasses. Journal of Applied Physics. 77: 1034-1039.

[28] Ogusu K, Kumagai T, Fujimori Y, Kitao M (2003) Thermal Analysis and Raman Scattering Study on Crystallization and Structure of $Ag_x(As_{0.4}Se_{0.6})_{100-x}$ Glasses. Journal of Non-Crystalline Solids. 324: 118-126.

[29] Saiter J M, Ledru J, Hamou A, Saffarini G (1998) Crystallization of As_xSe_{1-x} from the Glassy State (0.005<x<0.03). Physica B: Condensed Matter. 245: 256-262.

[30] Tonchev D, Fogal B, Belev G, Johanson R E, Kasap S O (2002) Properties of a-Sb_xSe_{1-x} Photoconductors. Journal of Non-Crystalline Solids. 299–302, Part 2: 998-1001.

[31] Mikla V I, Mikhalko I P, Mikla V V (2001) Laser-Induced Amorphous-to-Crystalline Phase Transition in Sb_xSe_{1-x} Alloys. Materials Science and Engineering: B. 83: 74-78.

[32] Holubová J, Černošek Z, Černošková E (2009) The Selenium Based Chalcogenide Glasses with Low Content of As and Sb: DSC, Stepscan DSC and Raman Spectroscopy Study. Journal of Non-Crystalline Solids. 355: 2050-2053.

[33] Savage J A, Webber P J, Pitt A M (1978) An Assessment of Ge-Sb-Se Glasses as 8 to 12μm Infra-Red Optical Materials. Journal of Materials Science. 13: 859-864.

[34] Giridhar A, Narasimham P S L, Mahadevan S (1980) Electrical Properties of Ge-Sb-Se Glasses. Journal of Non-Crystalline Solids. 37: 165-179.

[35] Shaaban E, Tomsah I (2011) The Effect of Sb Content on Glass-Forming Ability, the Thermal Stability, and Crystallization of Ge–Se Chalcogenide Glass. Journal of Thermal Analysis and Calorimetry. 105: 191-198.

[36] Peng Q, Dong Y, Deng Z, Kou H, Gao S, Li Y (2002) Selective Synthesis and Magnetic Properties of α-MnSe and MnSe$_2$ Uniform Microcrystals. The Journal of Physical Chemistry B. 106: 9261-9265.

[37] Vlaev L T, Tavlieva M P (2007) Structural and Thermal Studies on the Solid Products in the System MnSeO$_3$-SeO$_2$-H$_2$O. Journal of Thermal Analysis and Calorimetry. 90: 385-392.

[38] Vlaev L T, Genieva S D, Georgieva V G (2006) Study of the Crystallization Fields of Nickel(II) Selenites in the System NiSeO$_3$-SeO$_2$-H$_2$O. Journal of Thermal Analysis and Calorimetry. 86: 449-456.

[39] Gospodinov G G, Stancheva A G (2004) Physicochemical Study on Selenites of the Three-Component System Yb$_2$O$_3$-SeO$_2$-H$_2$O. Journal of Thermal Analysis and Calorimetry. 76: 537-542.

[40] Dammak M, Khemakhem H, Mhiri T (2001) Superprotonic Conduction and Ferroelectricity in Addition Cesium Sulfate Tellurate Cs$_2$SO$_4$.Te(OH)$_6$. Journal of Physics and Chemistry of Solids. 62: 2069-2074.

[41] Dammak M, Khemakhem H, Mhiri T, Kolsi A W, Daoud A (1999) Structural and Vibrational Study of K$_2$SeO$_4$·Te(OH)$_6$ Material. Journal of Solid State Chemistry. 145: 612-618.

[42] Dammak M, Khemakhem H, Mhiri T, Kolsi A W, Daoud A (1998) Structure and Characterization of a Mixed Crystal Rb$_2$SO$_4$.Te(OH)$_6$. Journal of Alloys and Compounds. 280: 107-113.

[43] Abdelhedi M, Dammak M, Cousson A, Kolsi A W (2005) Structural, Calorimetric and Conductivity Study of the New Mixed Solution Rb$_2$(SO$_4$)$_{0.5}$(SeO$_4$)$_{0.5}$Te(OH)$_6$. Journal of Alloys and Compounds. 398: 55-61.

[44] Ktari L, Abdelhedi M, Bouhlel N, Dammak M, Cousson A (2009) Synthesis, Calorimetric, Structural and Conductivity Studies in a New Thallium Selenate Tellurate Adduct Compound. Materials Research Bulletin. 44: 1792-1796.

[45] Pansuriya P, Parekh H, Friedrich H, Maguire G (In Press) Bisthiourea: Thermal and Structural Investigation. Journal of Thermal Analysis and Calorimetry. DOI: 10.1007/s10973-012-2309-3.

[46] Ramamurthi K, Madhurambal G, Ravindran B, Mariappan M, Mojumdar S (2011) The Growth and Characterization of a Metal Organic Crystal, Potassium Thiourea Thiocyanide. Journal of Thermal Analysis and Calorimetry. 104: 943-947.

[47] Rampon D S, Rodembusch F S, Schneider J M F M, Bechtold I H, Goncalves P F B, Merlo A A, Schneider P H (2010) Novel Selenoesters Fluorescent Liquid Crystalline Exhibiting a Rich Phase Polymorphism. Journal of Materials Chemistry. 20: 715-722.

[48] Xie Y, Zheng X, Jiang X, Lu J, Zhu L (2001) Sonochemical Synthesis and Mechanistic Study of Copper Selenides $Cu_{2-x}Se$, β-CuSe, and Cu_3Se_2. Inorganic Chemistry. 41: 387-392.

[49] Jiang Y, Xie B, Wu J, Yuan S, Wu Y, Huang H, Qian Y (2002) Room-Temperature Synthesis of Copper and Silver, Nanocrystalline Chalcogenides in Mixed Solvents. Journal of Solid State Chemistry. 167: 28-33.

[50] Chrissafis K, Paraskevopoulos K, Manolikas C (2006) Studying $Cu_{2-x}Se$ Phase Transformation through DSC Examination. Journal of Thermal Analysis and Calorimetry. 84: 195-199.

[51] Ribeiro da Silva M A V, Santos A F L O M, Gomes J R B, Roux M a V, Temprado M, Jiménez P, Notario R (2009) Thermochemistry of Bithiophenes and Thienyl Radicals. A Calorimetric and Computational Study. The Journal of Physical Chemistry A. 113: 11042-11050.

[52] Foces-Foces C, Roux M, Notario R, Segura M (2011) Thermal Behavior and Polymorphism in Medium–High Temperature Range of the Sulfur Containing Amino Acids L-Cysteine and L-Cystine. Journal of Thermal Analysis and Calorimetry. 105: 747-756.

[53] Kerr K A, Ashmore J P (1973) Structure and Conformation of Orthorhombic L-Cysteine. Acta Crystallographica Section B. 29: 2124-2127.

[54] Gorbitz C H, Dalhus B (1996) L-Cysteine, Monoclinic Form, Redetermination at 120K. Acta Crystallographica Section C. 52: 1756-1759.

[55] Kolesov B A, Minkov V S, Boldyreva E V, Drebushchak T N (2008) Phase Transitions in the Crystals of L- and DL-Cysteine on Cooling: Intermolecular Hydrogen Bonds Distortions and the Side-Chain Motions of Thiol-Groups. 1. L-Cysteine. The Journal of Physical Chemistry B. 112: 12827-12839.

[56] Moggach S A, Allan D R, Clark S J, Gutmann M J, Parsons S, Pulham C R, Sawyer L (2006) High-Pressure Polymorphism in L-Cysteine: The Crystal Structures of L-Cysteine-III and L-Cysteine-IV. Acta Crystallographica Section B. 62: 296-309.

[57] Oughton B M, Harrison P M (1959) The Crystal Structure of Hexagonal L-Cystine. Acta Crystallographica. 12: 396-404.

[58] Chaney M O, Steinrauf L K (1974) The Crystal and Molecular Structure of Tetragonal L-Cystine. Acta Crystallographica Section B. 30: 711-716.

[59] Dahaoui S, Pichon-Pesme V, Howard J A K, Lecomte C (1999) Ccd Charge Density Study on Crystals with Large Unit Cell Parameters: The Case of Hexagonal L-Cystine. The Journal of Physical Chemistry A. 103: 6240-6250.

[60] Regis Botelho J, Duarte Gondim A, Garcia dos Santos I, Dunstan P, Souza A, Fernandes V, Araújo A (2004) Thermochemical Parameters of Dimethyl and Di-Iso-Propyl Dithiocarbamate Complexes of Palladium(II). Journal of Thermal Analysis and Calorimetry. 75: 607-613.

[61] Airoldi C, Chagas A P (1992) Some Features of the Thermochemistry of Coordination-Compounds. Coordination Chemistry Reviews. 119: 29-65.

[62] de Souza A, Neto F, de Souza J, Macedo R, de Oliveira J, Pinheiro C (1997) Thermochemical Parameters of Complexes of Di-Isobutyldithiocarbamate with Phosphorus-Group Elements. Journal of Thermal Analysis and Calorimetry. 49: 679-684.

[63] Feng M-L, Xie Z-L, Huang X-Y (2009) Two Gallium Antimony Sulfides Built on a Novel Heterometallic Cluster. Inorganic Chemistry. 48: 3904-3906.

[64] Yao H-G, Ji M, Ji S-H, Zhang R-C, An Y-L, Ning G-l (2009) Solvothermal Syntheses of Two Novel Layered Quaternary Silver-Antimony(III) Sulfides with Different Strategies. Crystal Growth & Design. 9: 3821-3824.

[65] Ribeiro da Silva M A V, Santos A F L O M (2010) Thermochemical Properties of Two Nitrothiophene Derivatives 2-Acetyl-5-Nitrothiophene and 5-Nitro-2-Thiophenecarboxaldehyde. Journal of Thermal Analysis and Calorimetry. 100: 403-411.

[66] Morais C, Pat B, Gobe G, Johnson D W, Healy H (2006) Pyrrolidine Dithiocarbamate Exerts Anti-Proliferative and Pro-Apoptotic Effects in Renal Cell Carcinoma Cell Lines. Nephrology Dialysis Transplantation. 21: 3377-3388.

[67] Morais C, Gobe G, Johnson D W, Healy H (2009) Anti-Angiogenic Actions of Pyrrolidine Dithiocarbamate, a Nuclear Factor Kappa B Inhibitor. Angiogenesis. 12: 365-379.

[68] Kostova I (2006) Platinum Complexes as Anticancer Agents. Recent Patents on Anti-Cancer Drug Discovery. 1: 1-22.

[69] Yao X, Panichpisal K, Kurtzman N, Nugent K (2007) Cisplatin Nephrotoxicity: A Review. American Journal of the Medical Sciences. 334: 115-124.

[70] Galanski M, Arion V B, Jakupec M A, Keppler B K (2003) Recent Developments in the Field of Tumor-Inhibiting Metal Complexes. Current Pharmaceutical Design. 9: 2078-2089.

[71] Jones M M, Basinger M A, Mitchell W M, Bradley C A (1986) Inhibition of Cis-Diamminedichloroplatinum(II)-Induced Renal Toxicity in the Rat. Cancer Chemotherapy and Pharmacology. 17: 38-42.

[72] Gandara D R, Wiebe V J, Perez E A, Makuch R W, Degregorio M W (1990) Cisplatin Rescue Therapy - Experience with Sodium Thiosulfate, WR2721, and Diethyldithiocarbamate. Critical Reviews in Oncology/Hematology. 10: 353-365.

[73] Hidaka S, Tsuruoka M, Funakoshi T, Shimada H, Kiyozumi M, Kojima S (1994) Protective Effects of Dithiocarbamates against Renal Toxicity of Cis-Diamminedichloroplatinum in Rats. Renal Failure. 16: 337-349.

[74] Botelho J R, Gondim A D, Santos I M G, Dunstan P O, Souza A G, Fernandes V J, Araujo A S (2004) Thermochemical Parameters of Dimethyl and Di-Iso-Propyl Dithiocarbamate Complexes of Palladium(II). Journal of Thermal Analysis and Calorimetry. 75: 607-613.

[75] Botelho J, Souza A, Nunes L, Chagas A, Garcia dos Santos I, da Conceição M, Dunstan P (2002) Thermochemical Properties of Palladium(II) Chelates Involving Dialkyldithiocarbamates. Journal of Thermal Analysis and Calorimetry. 67: 413-417.

[76] Fregona D, Giovagnini L, Ronconi L, Marzano C, Trevisan A, Sitran S, Biondi B, Bordin F (2003) Pt(II) and Pd(II) Derivatives of Ter-Butylsarcosinedithiocarbamate. Synthesis, Chemical and Biological Characterization and in Vitro Nephrotoxicity. Journal of Inorganic Biochemistry. 93: 181-189.

[77] Giovagnini L, Marzano C, Bettio F, Fregona D (2005) Mixed Complexes of Pt(II) and Pd(II) with Ethylsarcosinedithiocarbamate and 2-/3-Picoline as Antitumor Agents. Journal of Inorganic Biochemistry. 99: 2139-2150.

[78] Faraglia G, Fregona D, Sitran S, Giovagnini L, Marzano C, Baccichetti F, Casellato U, Graziani R (2001) Platinum(II) and Palladium(II) Complexes with Dithiocarbamates and Amines: Synthesis, Characterization and Cell Assay. Journal of Inorganic Biochemistry. 83: 31-40.

[79] Manav N, Mishra A K, Kaushik N K (2006) In Vitro Antitumour and Antibacterial Studies of Some Pt(IV) Dithiocarbamate Complexes. Spectrochimica Acta Part A-Molecular and Biomolecular Spectroscopy. 65: 32-35.

[80] Manav N, Mishra A K, Kaushik N K (2004) Triphenyl Phosphine Adducts of Platinum(IV) and Palladium(II) Dithiocarbamates Complexes: A Spectral and in Vitro Study. Spectrochimica Acta Part A-Molecular and Biomolecular Spectroscopy. 60: 3087-3092.

[81] Uivarosi V, Badea M, Aldea V, Chirigiu L, Olar R (In Press) Thermal and Spectral Studies of Palladium(II) and Platinum(IV) Complexes with Dithiocarbamate Derivatives. Journal of Thermal Analysis and Calorimetry.

[82] Rademaker-Lakhai J M, van den Bongard D, Pluim D, Beijnen J H, Schellens J H M (2004) A Phase I and Pharmacological Study with Imidazolium-Trans-Dmso-Imidazole-Tetrachlororuthenate, a Novel Ruthenium Anticancer Agent. Clinical Cancer Research. 10: 3717-3727.

[83] Alessio E, Mestroni G, Nardin G, Attia W M, Calligaris M, Sava G, Zorzet S (1988) Cis-Dihalotetrakis(Dimethyl Sulfoxide)Ruthenium(II) and Trans-Dihalotetrakis(Dimethyl Sulfoxide)Ruthenium(II) Complexes $(RuX_2(Dmso)_4, X = Cl, Br)$ - Synthesis, Structure, and Antitumor-Activity. Inorganic Chemistry. 27: 4099-4106.

[84] Alessio E, Iengo E, Geremia S, Calligaris M (2003) New Geometrical and Linkage Isomers of the Ru(II) Precursor $Cis,Cis,Trans$-$RuCl_2(Dmso$-$S)_2(Dmso$-$O)(CO)$: A Spectroscopic and Structural Investigation. Inorganica Chimica Acta. 344: 183-189.

[85] Zhelev Z, Ohba H, Bakalova R, Hadjimitova V, Ishikawa M, Shinohara Y, Baba Y (2004) Phenothiazines Suppress Proliferation and Induce Apoptosis in Cultured Leukemic Cells without Any Influence on the Viability of Normal Lymphocytes - Phenothiazines and Leukemia. Cancer Chemotherapy and Pharmacology. 53: 267-275.

[86] Hollo B, Krstic M, Sovilj S P, Pokol G, Szecsenyi K M (2011) Thermal Decomposition of New Ruthenium(II) Complexes Containing N-Alkylphenothiazines. Journal of Thermal Analysis and Calorimetry. 105: 27-32.

[87] Plano D, Sanmartin C, Moreno E, Prior C, Calvo A, Palop J A (2007) Novel Potent Organoselenium Compounds as Cytotoxic Agents in Prostate Cancer Cells. Bioorganic & Medicinal Chemistry Letters. 17: 6853-6859.

[88] Plano D, Baquedano Y, Ibáñez E, Jiménez I, Palop J A, Spallholz J E, Sanmartín C (2010) Antioxidant-Prooxidant Properties of a New Organoselenium Compound Library. Molecules. 15: 7292-7312.

[89] Ibáñez E, Plano D, Font M, Calvo A, Prior C, Palop J A, Sanmartín C (2011) Synthesis and Antiproliferative Activity of Novel Symmetrical Alkylthio- and Alkylseleno-Imidocarbamates. European Journal of Medicinal Chemistry. 46: 265-274.

[90] Ibáñez E, Agliano A, Prior C, Nguewa P, Redrado M, González-Zubeldia I, Plano D, Palop J A, Sanmartin C, Calvo A (2012) The Quinoline Imidoselenocarbamate EI201 Blocks the Akt/Mtor Pathway and Targets Cancer Stem Cells Leading to a Strong Antitumor Activity. Current Medicinal Chemistry. 19: 3031-3043.

[91] Plano D, Lizarraga E, Font M, Palop J A, Sanmartin C (2009) Thermal Stability and Decomposition of Sulphur and Selenium Compounds. Journal of Thermal Analysis and Calorimetry. 98: 559-566.

[92] Kitamura S, Miyamae A, Koda S, Morimoto Y (1989) Effect of Grinding on the Solid-State Stability of Cefixime Trihydrate. International Journal of Pharmaceutics. 56: 125-134.

[93] Perrenot B, Widmann G (1994) Polymorphism by Differential Scanning Calorimetry. Thermochimica Acta. 234: 31-39.

[94] Plano D, Lizarraga E, Palop J, Sanmartín C (2011) Study of Polymorphism of Organosulfur and Organoselenium Compounds. Journal of Thermal Analysis and Calorimetry. 105: 1007-1013.

[95] Sanmartin C, Plano D, Dominguez E, Font M, Calvo A, Prior C, Encio I, Palop J A (2009) Synthesis and Pharmacological Screening of Several Aroyl and Heteroaroyl Selenylacetic Acid Derivatives as Cytotoxic and Antiproliferative Agents. Molecules. 14: 3313-3338.

Chapter 8

Calorimetric Determination of Heat Capacity, Entropy and Enthalpy of Mixed Oxides in the System CaO–SrO–Bi₂O₃–Nb₂O₅–Ta₂O₅

Jindřich Leitner, David Sedmidubský, Květoslav Růžička and Pavel Svoboda

Additional information is available at the end of the chapter

1. Introduction

Mixed oxides in the system CaO–SrO–Bi₂O₃–Nb₂O₅–Ta₂O₅ possess many extraordinary electric, magnetic and optical properties for which they are used in fabrication of various electronic components. For example $Sr_2(Nb,Ta)_2O_7$ and $(Sr,Ca)Bi_2(Nb,Ta)_2O_9$ are used for ferroelectric memory devices, $CaNb_2O_6$, $Sr_5(Nb_{1-x}Ta_x)_4O_{15}$ and $Bi(Nb,Ta)O_4$ for microwave dielectric resonators and $Ca_2Nb_2O_7$ as non-linear optical materials and hosts for rare-earth ions in solid-state lasers. Ternary strontium bismuth oxides $SrBi_2O_4$, $Sr_2Bi_2O_5$, and $Sr_6Bi_2O_9$ are of considerable interest due to a visible light driven fotocatalytic activity.

To assess the thermodynamic stability and reactivity of these oxides under various conditions during their preparation, processing and operation, a complete set of consistent thermodynamic data, including heat capacity, entropy and enthalpy of formation, is necessary. Some of these data are available in literature. Akishige et al. [1] have been measured the heat capacities of $Sr_2Nb_2O_7$ and $Sr_2Ta_2O_7$ single crystals in the temperature range 2-600 K. The results have been only plotted and the values of $S°_m(298)$ have not been calculated. A commensurate transformation of $Sr_2Nb_2O_7$ at T_{INC} = 495 K has been observed accompanied by changes in enthalpy and entropy of ΔH = 291 J mol⁻¹ and ΔS = 0.587 J K⁻¹ mol⁻¹. The heat capacity of $Sr_2Nb_2O_7$ has been also measured by Shabbir at al. [2] in the temperature range 375-575 K. They have observed a phase transition at T_{INC} = 487 ± 2 K connected with ΔH = 147 ± 14 J mol⁻¹ and ΔS = 0.71 ± 0.10 J K⁻¹ mol⁻¹. The heat capacities of polycrystalline and monocrystalline $SrBi_2Ta_2O_9$ and $Sr_{0,85}Bi_{2,1}Ta_2O_9$ have been measured by Onodera at al. [3–5] at 80-800 K. Morimoto at al. [6] have reported the results of the heat capacity measurements of $SrBi_2(Nb_xTa_{1-x})_2O_9$ (x = 0, 1/3, 2/3 a 1). The temperature dependences of heat capacities show lambda-transitions with maxima at the Currie temperature T_C = 570 ± 1 K, 585 ± 2 K, 625 ± 3 K a 690 ± 2 K for x = 0, 1/3, 2/3 and 1, respectively. Using EMF (electromotive force) measurements, Raghavan has obtained the values of the Gibbs energy of formation from binary oxides, $\Delta_{ox}G$, for some niobates [7,8] and tantalates [9,10] of calcium. His results are summarized in Table 1. The same technique has been employed by Dneprova et al. [11] for $\Delta_{ox}G$ measurement for $CaNb_2O_6$ and $Ca_2Nb_2O_7$. Their results presented in Table 1 are not significantly different from the results of Raghavan. Using the CALPHAD approach [13], Yang et al. [14] have assessed thermodynamic data for various mixed oxides in the $SrO-Nb_2O_5$ system. The same approach has been used by Hallstedt et al. for the assessment of thermodynamic data of mixed oxides in the systems $CaO-Bi_2O_3$ [14] and $SrO-Bi_2O_3$ [15]. Besides equilibrium data, values of the enthalpy of formation [16] of mixed oxide have been considered. Later on, these systems have been studied by EMF method by Jacob and Jayadevan [17,18] and temperature dependences of $\Delta_{ox}G$ for various mixed oxides have been derived. These data have been included into the thermodynamic re-assessment of the $CaO-SrO-Bi_2O_3$ system [19].

Oxide	$\Delta_{ox}G$ (kJ mol^{-1})	T (K)	$\Delta_{ox}H$ (kJ mol^{-1})	$\Delta_{ox}S$ (J K^{-1} mol^{-1})	Ref.
$CaNb_2O_6$	$-75.82 - 0.03345T$	1245-1300	-75.82	33.45	[7]
$Ca_2Nb_2O_7$	-178.44	1256			[8]
$Ca_3Nb_2O_8$	-209.94	1256			[8]
$CaTa_4O_{11}$	$-36.982 - 0.029T$	1250-1300	-36.98	29.0	[9]
$CaTa_2O_6$	-65.14	1250			[10]
$Ca_2Ta_2O_7$	-102.82	1250			[10]
$Ca_4Ta_2O_9$	-165.05	1250			[10]
$CaNb_2O_6$	$-175.73 + 0.02259T$	1100-1276	-175.73	-22.59	[11]
$Ca_2Nb_2O_7$	$-212.54 - 0.02218T$	1100-1350	-212.54	22.18	[11]
$Sr_2Nb_{10}O_{27}$	$-1125.69 + 0.35069T$	298-5000	-1125.69	-350.69	[12]
$SrNb_2O_6$	$-325.04 + 0.05865T$	298-5000	-325.04	-58.65	[12]
$Sr_2Nb_2O_7$	$-367.43 + 0.03993T$	298-5000	-367.43	-39.93	[12]
$Sr_5Nb_4O_{14}$	$-746.72 + 0.05101T$	298-5000	-746.72	-51.01	[12]
$Ca_5Bi_{14}O_{26}$	$-125.90 - 0.055T$	298-1300	-125.9	55.0	[19]
$CaBi_2O_4$	$-27.60 - 0.003T$	298-1300	-27.6	3.0	[19]
$Ca_4Bi_6O_{13}$	$-97.60 - 0.008T$	298-1300	-97.6	8.0	[19]
$Ca_2Bi_2O_5$	$-42.20 - 0.003T$	298-1300	-42.2	3.0	[19]
$SrBi_2O_4$	$-63.86 - 0.0018T$	298-1300	-63.86	1.8	[19]
$Sr_2Bi_2O_5$	$-118.75 + 0.024T$	298-1300	-118.75	-24.0	[19]
$Sr_3Bi_2O_6$	$-109.60 + 0.0024T$	298-1300	-109.60	-2.4	[19]

Table 1. Published values of $\Delta_{ox}G$, $\Delta_{ox}H$ a $\Delta_{ox}S$ for some mixed oxides in the system $CaO-SrO-Bi_2O_3-Nb_2O_5-Ta_2O_5$

This review brings a summary of our results [20–30] focused on calorimetric determination of heat capacity, entropy end enthalpy of mixed oxides in the system CaO–SrO–Bi$_2$O$_3$–Nb$_2$O$_5$–Ta$_2$O$_5$. Temperature dependences of molar heat capacity in a broad temperature range were evaluated from the experimental heat capacity and relative enthalpy data. Molar entropies at T = 298.15 K were calculated from low temperature heat capacity measurements. Furthermore, the results of calorimetric measurements of the enthalpies of drop-solution in a sodium oxide-molybdenum oxide melt for several stoichiometric mixed oxides in the above mentioned system are reported from which the values of enthalpy of formation from constituent binary oxides were derived. Finally, some empirical estimation and correlation methods (the Neumann-Kopp's rule, entropy-volume correlation and electronegativity-differences method) for evaluation of thermodynamic data of mixed oxides are tested and assessed.

2. Experimental

Nineteen mixed oxides in the system CaO–SrO–Bi$_2$O$_3$–Nb$_2$O$_5$–Ta$_2$O$_5$ with stoichiometry CaBi$_2$O$_4$, Ca$_4$Bi$_6$O$_{13}$, Ca$_2$Bi$_2$O$_5$, SrBi$_2$O$_4$, Sr$_2$Bi$_2$O$_5$, CaNb$_2$O$_6$, Ca$_2$Nb$_2$O$_7$, SrNb$_2$O$_6$, Sr$_2$Nb$_2$O$_7$, Sr$_2$Nb$_{10}$O$_{27}$, Sr$_5$Nb$_4$O$_{15}$, BiNbO$_4$, BiNb$_5$O$_{14}$, BiTaO$_4$, Bi$_4$Ta$_2$O$_{11}$, Bi$_7$Ta$_3$O$_{18}$, Bi$_3$TaO$_7$, SrBi$_2$Nb$_2$O$_9$, and SrBi$_2$Ta$_2$O$_9$ were prepared, characterized and examined. The samples were prepared by conventional solid state reactions from high purity precursors (CaCO$_3$, SrCO$_3$ Bi$_2$O$_3$, Nb$_2$O$_5$ and Ta$_2$O$_5$). A three step procedure was used consisting of an initial calcination run of mixed powder precursors and subsequent double firing of prereacted mixtures pressed into pellets. The phase composition of the prepared samples was checked by X-ray powder diffraction (XRD). XRD data were collected at room temperature with an X'Pert PRO (PANalytical, the Netherlands) θ-θ powder diffractometer with parafocusing Bragg-Brentano geometry using CuK$_\alpha$ radiation (λ = 1.5418 nm). Data were scanned over the angular range 5–60° (2θ) with an increment of 0.02° (2θ) and a counting time of 0.3 s step^{-1}. Data evaluation was performed by means of the HighScore Plus software.

The PPMS equipment 14 T-type (Quantum Design, USA) was used for the heat capacity measurements in the low temperature region [31-35]. The measurements were performed by the relaxation method [36] with fully automatic procedure under high vacuum (pressure ~10^{-2} Pa) to avoid heat loss through the exchange gas. The samples were compressed powder pellets. The densities of the samples were about 65 % of the theoretical ones.

The samples were mounted to the calorimeter platform with cryogenic grease Apiezon N (supplied by Quantum Design). The procedure was as follows: First, a blank sample holder with the Apiezon only was measured in the temperature range approx. 2–280 K to obtain background data, then the sample plate was attached to the calorimeter platform and the measurement was repeated in the same temperature range with the same temperature steps. The sample heat capacity was then obtained as a difference between the two data sets. This procedure was applied, because the heat capacity of Apiezon is not negligible in comparison with the sample heat capacity (~8 % at room temperature) and exhibits a peak-shaped transition below room temperature [37]. The manufacturer claims the precision of this measurement better then 2 % [38]; the control measurement of the copper sample (99.999 % purity) confirmed this precision in the temperature range 50–250 K. However, the precision of the measurement strongly depends on the thermal coupling between the sample and the calorimeter platform. Due to unavoidable porosity of the sample plate this coupling is rapidly getting worse as the temperature raises above 270 K and Apiezon diffuses into the porous sample. Consequently, the uncertainty of the obtained data tends to be larger.

A Micro DSC III calorimeter (Setaram, France) was used for the heat capacity determination in the temperature range of 253–352 K. First, the samples were preheated in a continuous mode from room temperature up to 352 K (heating rate 0.5 K min^{-1}). Then the heat capacity was measured in the incremental temperature scanning mode consisting of a number of 5–10 K steps (heating rate 0.2 K min^{-1}) followed by isothermal delays of 9000 s. Two subsequent step-by-step heating were recorded for each sample. Synthetic sapphire, NIST Standard reference material No. 720, was used as the reference. The uncertainty of heat capacity measurements is estimated to be better than ±1 %.

Enthalpy increment determinations were carried out by drop method using high-temperature calorimeter, Multi HTC 96 (Setaram, France). All measurements were performed in air by alternating dropping of the reference material (small pieces of synthetic sapphire, NIST Standard reference material No. 720) and of the sample (pressed pellets 5 mm in diameter) being initially held at room temperature, through a lock into the working cell of the preheated calorimeter. Endothermic effects are detected and the relevant peak area is proportional to the heat content of the dropped specimen. The delays between two subsequent drops were 25–30 min. To check the accuracy of measurement, the enthalpy increments of platinum in the temperature range 770–1370 K were measured first and compared with published reference values [39]. The standard deviation of 22 runs was 0.47 kJ mol^{-1}, the average relative error was 2.0 %. Estimated overall accuracy of the drop measurements is ±3 %.

The heats of drop-solution were determined using a Multi HTC 96 high-temperature calorimeter (Setaram, France). A sodium oxide-molybdenum oxide melt of the stoichiometry $3Na_2O + 4MoO_3$ was used as the solvent. The ratio of solute/solvent varied from 1/250 up to 1/500. The measurements were performed at temperatures of 973 and 1073 K in argon or air atmosphere. The method consists in alternating dropping of the reference material (small spherules of pure platinum) and of the sample (small pieces of pressed tablets 10–40 mg), being initially held near room temperature (T_0), through a lock into the working cell (a platinum crucible with the solvent) of the preheated calorimeter at temperature T. Two or three samples were examined during one experimental run. The delays between two subsequent drops were 30–60 min. The total heat effect ($\Delta_{ds}H$) includes the heat of solution ($\Delta_{sol}H$), the heat content of the sample ($\Delta_T H$), and, for the carbonates, the heat of decomposition ($\Delta_{decomp}H$) to form solid CaO or SrO and gaseous CO_2. Using appropriate thermochemical cycles, the values of the enthalpy of formation of mixed oxides from the binary oxides and from the elements at 298 K were evaluated. The temperature dependence of the heat capacity of platinum [39] was used for the calculation of the sensitivity of the calorimeters.

2.1. Characterization of prepared samples

The XRD analysis revealed that the prepared samples were without any observable diffraction lines from unreacted precursors or other phases. The lattice parameters of the oxides were evaluated by Rietveld refinement [40] and are summarized in Table 2 together with the values of theoretical density calculated from the lattice parameters.

2.2. Evaluation of temperature dependence of heat capacity at low temperatures

The fit of the low-temperature heat capacity data (LT fit) consists of two steps. Assuming the validity of the phenomenological formula $C_{pm} = \beta T^3 + \gamma_{el}T$, at $T \rightarrow 0$ where β is proportional to the inverse cube root of the Debye temperature Θ_D and $\gamma_{el}T$ is the Sommerfeld term, we plotted the C_{pm}/T vs. T^2 dependence for $T < 8$ K to estimate the Θ_D and γ_{el} values. Since all compounds under study are semiconductors with a sufficiently large band gap, the non-zero γ_{el} values are supposed to be either due to some metallic impurities or to a series of Schottky-like transitions resulting from structure defects. Nevertheless, they are negligible in most cases (typically < 0.5 mJ K^{-2} mol^{-1}) and can be ignored in further analysis. As an example, the results of heat capacity measurements on $CaNb_2O_6$ and LT fit for $T < 10$ K is shown in Fig. 1.

Oxide	a (nm)	b (nm)	c (nm)	α (°)	β (°)	γ (°)	d (g cm⁻³)	Ref.
CaBi₂O₄	1.66143	1.15781	1.39915	90	134.03	90	6.631	[20]
Ca₄Bi₆O₁₃	0.59308	1.73512	0.72192	90	90	90	6.540	[20]
Ca₂Bi₂O₅	1.01074	1.01249	1.04618	116.88	107.16	92.98	6.468	[20]
SrBi₂O₄	1.92635	0.43437	0.61444	90	95.50	90	7.392	[29]
Sr₂Bi₂O₅	1.42935	0.61715	0.76478	90	90	90	6.628	[29]
CaNb₂O₆	1.49698	0.57472	0.52202	90	90	90	4.760	[26]
Ca₂Nb₂O₇	0.76853	1.33587	0.54959	90	90	98.29	4.496	[26]
SrNb₂O₆	0.77209	0.55930	1.09821	90	90.37	90	5.174	[24]
Sr₂Nb₂O₇	0.39544	2.67735	0.57004	90	90	90	5.206	[26]
Sr₂Nb₁₀O₂₇	3.715	3.697	0.3943	90	90	90	5.653	a)
Sr₅Nb₄O₁₅	0.56576	0.56576	1.14536	90	90	120	5.490	[27]
BiNbO₄	0.56893	1.1728	0.49915	90	90	90	7.297	[21]
BiNb₅O₁₄	1.76762	1.72072	0.39610	90	90	90	4.948	b)
BiTaO₄	0.56394	1.1776	0.49626	90	90	90	9.149	[21]
Bi₄Ta₂O₁₁	0.66159	0.76528	0.98781	101.39	90.10	89.99	9.306	[28]
Bi₇Ta₃O₁₈	3.40162	0.76054	0.66354	90	109.16	90	9.395	[28]
Bi₃TaO₇	0.54711	0.54711	0.54711	90	90	90	9.327	[28]
SrBi₂Nb₂O₉	0.55160	0.55087	2.51020	90	90	90	7.275	[22]
SrBi₂Ta₂O₉	0.55224	0.55266	2.50124	90	90	90	8.801	[22]

a) Quoted according to JCPDS 035-1220.
b) Quoted according to JCPDS 048-0986

Table 2. Structural characterization of prepared samples

In the second step of the LT fit, both sets of the C_{pm} data (relaxation time + DSC) were considered. Analysis of the phonon heat capacity was performed as an additive combination of Debye and Einstein models. Both models include corrections for anharmonicity, which is responsible for a small, but not negligible, additive term at higher temperatures and which accounts for the difference between isobaric and isochoric heat capacity. According to literature [41], the term $1/(1 - \alpha T)$ is considered as a correction factor.

The acoustic part of the phonon heat capacity is described using the Debye model

$$C_{phD} = \frac{9R}{1 - \alpha_D T} \left(\frac{T}{\Theta_D} \right)^3 \int_0^{x_D} \frac{x^4 \exp(x)}{\left[\exp(x) - 1 \right]^2} dx \tag{1}$$

where R is the gas constant, Θ_D is the Debye characteristic temperature, α_D is the coefficient of anharmonicity of acoustic branches and $x_D = \Theta_D/T$. Here the three acoustic branches are taken as one triply degenerate branch. Similarly, the individual optical branches are described by the Einstein model

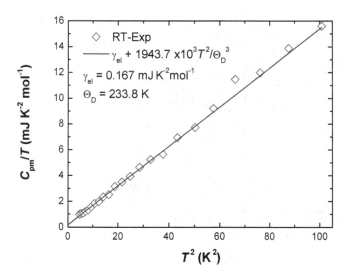

Figure 1. Temperature dependence of C_{pm}/T function for CaNb$_2$O$_7$ at low temperatures

$$C_{phEi} = \frac{R}{1 - \alpha_{Ei} T} \frac{x_{Ei}^2 \exp(x_{Ei})}{\left[\exp(x_{Ei}) - 1 \right]^2} \tag{2}$$

where α_{Ei} and $x_{Ei} = \Theta_{Ei}/T$ have analogous meanings as in the previous case. Several optical branches are again grouped into one degenerate multiple branch with the same Einstein characteristic temperature and anharmonicity coefficient. The phonon heat capacity then reads

Calorimetric Determination of Heat Capacity, Entropy and Enthalpy of Mixed Oxides in the System
CaO–SrO–Bi$_2$O$_3$–Nb$_2$O$_5$–Ta$_2$O$_5$

173

$$C_{ph} = C_{phD} + \sum_{i=1}^{3n-3} C_{phEi} \tag{3}$$

All the estimated values were further treated by a simplex routine and a full non-linear fit was performed on all adjustable parameters.

The values of relative enthalpies at 298.15 K, $H_m(298.15) - H_m(0)$, were evaluated from the low-temperature C_{pm} data (LT fit) by numerical integration of the $C_{pm}(T)$ dependences from zero to 298.15 K. Standard deviations (2σ) were calculated using the error propagation law. The values of standard molar entropies at 298.15 K, $S_m(298.15)$, were derived from the low-temperature C_{pm} data (LT fit) by numerical integration of the $C_{pm}(T)/T$ dependences from zero to 298.15 K. A numerical integration was used with the boundary conditions $S_m = 0$ and $C_{pm} = 0$ at $T = 0$ K. Standard deviations (2σ) were calculated using the error propagation law. All calculated values are summarized in Table 3.

Oxide	$C_{pm}(298)$ (J K^{-1} mol^{-1})	$H_m(298)-H_m(0)$ (J mol^{-1})	$S_m(298)$ (J K^{-1} mol^{-1})	$\Delta_{ox}S$ (J K^{-1} mol^{-1})	Ref.
CaBi$_2$O$_4$	151.3	26470 ± 158	188.5 ± 3.3	1.9	[20]
Ca$_4$Bi$_6$O$_{13}$	504.1	85079 ± 507	574.1 ± 8.8	−23.8	[20]
Ca$_2$Bi$_2$O$_5$	197.4	33735 ± 201	231.3 ± 2.9	6.6	[20]
SrBi$_2$O$_4$	155.6	29601 ± 169	206.1 ± 1.1	4.0	[29]
Sr$_2$Bi$_2$O$_5$	201.9	38199 ± 219	261.2 ± 1.4	5.5	[29]
CaNb$_2$O$_6$	171.8	28159 ± 170	167.3 ± 0.9	−8.1	[26]
Ca$_2$Nb$_2$O$_7$	218.1	35631 ± 215	212.4 ± 1.2	−1.1	[26]
SrNb$_2$O$_6$	170.2	28722 ± 174	173.9 ± 0.9	−17.0	[24]
Sr$_2$Nb$_2$O$_7$	216.6	37977 ± 266	238.5 ± 1.3	−5.9	[26]
Sr$_2$Nb$_{10}$O$_{27}$	746.8	124150 ± 740	759.7 ± 4.1	−33.9	[27]
Sr$_5$Nb$_4$O$_{15}$	477.2	83340 ± 490	524.5 ± 2.8	−18.4	[27]
BiNbO$_4$	121.3	22120 ± 134	147.9 ± 0.8	5.0	[21]
BiNb$_5$O$_{14}$	386.8	62639 ± 362	397.2 ± 2.1	−25.8	[23]
BiTaO$_4$	119.3	22021 ± 132	149.1 ± 0.8	3.3	[21]
Bi$_4$Ta$_2$O$_{11}$	363.2	66566 ± 384	449.6 ± 2.3	9.5	[28]
Bi$_7$Ta$_3$O$_{18}$	602.7	109760 ± 634	743.0 ± 3.8	8.6	[28]
Bi$_3$TaO$_7$	235.2	44265 ± 254	304.3 ± 1.6	10.0	[28]
SrBi$_2$Nb$_2$O$_9$	286.4	49230 ± 292	327.2 ± 1.7	−12.2	[22]
SrBi$_2$Ta$_2$O$_9$	286.6	49060 ± 289	339.2 ± 1.8	−5.9	[22]

Table 3. Heat capacity, relative enthalpy, entropy and entropy of formation from binary oxides at temperature 298.15 K of various mixed oxides in the system CaO–SrO–Bi$_2$O$_3$–Nb$_2$O$_5$–Ta$_2$O$_5$

A comparison is given in Table 4 of the values of entropy of formation from binary oxides $\Delta_{ox}S$ at 298 K calculated from our results and those from literature. The values of $\Delta_{ox}S$ are calculated using the relation

$$\Delta_{ox}S = S_m(MO) - \sum_i b_i\, S_m(BO,i) \tag{4}$$

where $S_m(MO)$ and $S_m(BO,i)$ stand for the molar entropies of a mixed oxide and a binary oxide i, respectively, and b_i is a constitution coefficient representing the number of formula units of a binary oxide i per formula unit of the mixed oxide. The following values were used for calculation: $S_m(CaO,298.15\ K) = 38.1\ J\ K^{-1}\ mol^{-1}$ [42], $S_m(SrO,298.15\ K) = 53.58\ J\ K^{-1}\ mol^{-1}$ [43], $S_m(Bi_2O_3,298.15\ K) = 148.5\ J\ K^{-1}\ mol^{-1}$ [44], $S_m(Nb_2O_5,\ 298.15) = 137.30\ J\ K^{-1}\ mol^{-1}$ [45] $S_m(Ta_2O_5,\ 298.15) = 143.09\ J\ K^{-1}\ mol^{-1}$ [45]. Furthermore, $S_m(Sr_2Nb_2O_7,\ 298.15) = 238.5\ J\ K^{-1}\ mol^{-1}$ from this work can be directly compared with the value $232.37\ J\ K^{-1}\ mol^{-1}$ obtained by numeric integration of the $C_{pm}(T)/T$ dependences from zero to 298.15 K given in Ref. [1]. It should be noted that the values of entropy assessed by thermodynamic optimization of phase equilibrium data are generally considered as less reliable as the values derived from low temperature heat capacity measurements. It is due to possible strong correlation between the enthalpy and entropy contributions to the Gibbs energy. So the obvious discrepancies between our values and data from assessments [12,19] could be explain in this way.

Oxide	$\Delta_{ox}S$ [a) $(J\ K^{-1}\ mol^{-1})$	$\Delta_{ox}S$ $(J\ K^{-1}\ mol^{-1})$	Ref.
CaNb_2O_6	−8.1	33.45	[7]
		−22.59	[11]
Ca_2Nb_2O_7	−1.1	22.18	[11]
Sr_2Nb_10O_27	−34.0	−350.69	[12]
SrNb_2O_6	−17.0	−58.65	[12]
Sr_2Nb_2O_7	−6.0	−39.93	[12]
Sr_5Nb_4O_14	−18.0	−51.01	[12]
CaBi_2O_4	1.9	3.0	[19]
Ca_4Bi_6O_13	−23.8	8.0	[19]
Ca_2Bi_2O_5	6.6	3.0	[19]
SrBi_2O_4	4.0	1.8	[19]
Sr_2Bi_2O_5	5.5	−24.0	[19]

a) This work

Table 4. The values of entropy of formation from binary oxides at 298.15 K: a comparison of our results and data from literature

It should be noted that the thorough analysis of the Debye and Einstein contributions to the heat capacities reveals that the different vibrational modes contribute to the total values of $\Delta_{ox}S$ to a different extent and partial compensation is possible in some cases.

2.3. Evaluation of heat capacity at temperatures above 298 K

For the assessment of temperature dependences of C_{pm} above room temperature, the heat capacity data from DSC and the enthalpy increment data from drop calorimetry were treated simultaneously by the linear least-squares method (HT fit). The temperature dependence of C_{pm} was considered in the form

$$C_{pm} = A + BT + C/T^2 \tag{5}$$

thus the related temperature dependence of $\Delta H_m(T) = H_m(T) - H_m(T_0)$ is given by equation

$$\Delta H_m(T) = H_m(T) - H_m(T_0) = \int_{T_0}^{T} C_{pm} dT = A(T - T_0) + B(T^2 - T_0^2)/2 - C(1/T - 1/T_0) \tag{6}$$

The sum of squares which is minimized has the following form

$$F = \sum_{i=1}^{N(C_p)} w_i^2 \left[C_{pm,i} - A - BT_i - C/T_i^2 \right]^2 + $$
$$+ \sum_{j=1}^{N(\Delta H)} w_j^2 \left[\Delta H_{m,j} - A(T_j - T_{0,j}) - B(T_j^2 - T_{0,j}^2)/2 + C(1/T_j - 1/T_{0,j}) \right]^2 \rightarrow \min \tag{7}$$

where the first sum runs over the C_{pm} experimental points while the second sum runs over the ΔH_m experimental points. Different weights w_i (w_j) were assigned to individual points calculated as $w_i = 1/\delta_i$ ($w_j = 1/\delta_j$) where δ_i (δ_j) is the absolute deviation of the measurement estimated from overall accuracies of measurements (1 % for DSC and 3 % for drop calorimetry). Both types of experimental data thus gain comparable significance during the regression analysis. To smoothly connect the LT fit and HT fit data the values of $C_{pm}(298.15)$ from LT fit were used as constraints and so Eq. (7) is modified

$$F_{constr} = F - \lambda \left[C_{pm}(298.15) - A - 298.15B - C/298.15^2 \right] \rightarrow \min \tag{8}$$

The numerical values of parameters A, B and C are now obtained by solving a set of equations deduced as derivatives of F_{constr} with respect of these parameters and a multiplier λ which are equal to zero at the minimum of F_{constr}. Assessed values of parameters A, B and C of Eq. (4) for mixed oxides are presented in Table 5.

As an example, the results of heat capacity measurements and relative enthalpy measurements on Bi$_7$Ta$_3$O$_{18}$ [28] are shown in Fig. 2. Empirical estimation according to the Neumann-Kopp's rule (NKR) is also plotted for comparison.

The empirical Neumann-Kopp's rule (NKR) is frequently used for estimation of unknown values of the heat capacity of mixed oxides [46–48]. According to NKR, heat capacity of a mixed oxide is calculated as a sum of heat capacities of the constituent binary ones

$$C_{pm}(MO) = \sum_i b_i C_{pm}(BO,i) \tag{9}$$

It was concluded [47,48] that NKR predicts the heat capacities of mixed oxides remarkably well around room temperature but the deviations (mostly positive) from NKR become substantial at higher temperatures. Mean relative error of the estimated values of $C_{pm}(298.15\ K)$ is 1.4 %. Calculated temperature dependences of $\Delta_{ox}C_{pm} = C_{pm}(MO) - \Sigma b_i C_{pm}(BO,i)$ for various mixed oxides in the systems $CaO–Nb_2O_5$, $SrO–Nb_2O_5$ and $Bi_2O_3–Ta_2O_5$ are shown in Fig. 3.

Oxide	$C_{pm} = A + B \cdot T + C/T^2$ (J K^{-1} mol^{-1})			Temperature range (K)	Ref.
	A	10^3 B	10^{-6} C		
$CaBi_2O_4$	157.161	38.750	−1.546	298-1000	[20]
$Ca_4Bi_6O_{13}$	550.808	114.890	−7.201	298-1200	[20]
$Ca_2Bi_2O_5$	226.096	33.374	−3.432	298-1100	[20]
$SrBi_2O_4$	161.97	45.936	−1.7832	298–1100	[29]
$Sr_2Bi_2O_5$	197.48	87.463	−1.9282	298–1200	[29]
$CaNb_2O_6$	200.40	34.32	−3.45	298-1500	[26]
$Ca_2Nb_2O_7$	257.20	36.21	−4.435	298-1400	[26]
$SrNb_2O_6$	200.47	29.37	−3.473	298-1500	[24]
$Sr_2Nb_2O_7$	248.00	43.50	−3.948	298-1400	[26]
$Sr_2Nb_{10}O_{27}$	835.351	227.648	−13.904	298-1400	[27]
$Sr_5Nb_4O_{15}$	504.796	147.981	−6.376	298-1400	[27]
$BiNbO_4$ [a]	128.628	33.400	−1.991	150-1200	[21]
$BiNb_5O_{14}$	455.840	60.160	−7.734	298-1400	[23]
$BiTaO_4$ [b]	133.594	25.390	−2.734	150-1200	[21]
$Bi_4Ta_2O_{11}$	445.8	5.451	−7.489	298–1400	[28]
$Bi_7Ta_3O_{18}$	699.0	52.762	−9.956	298–1400	[28]
Bi_3TaO_7	251.6	67.05	−3.237	298-1400	[28]
$SrBi_2Nb_2O_9$	324.470	63.710	−5.076	298-1400	[22]
$SrBi_2Ta_2O_9$	320.220	64.510	−4.700	298-1400	[22]

[a] An extra term $1.363 \times 10^8/T^3$ was added.
[b] An extra term $2.360 \times 10^8/T^3$ was added.

Table 5. Parameters of temperature dependence of molar heat capacities of various mixed oxides in the system $CaO–SrO–Bi_2O_3–Nb_2O_5–Ta_2O_5$

Calorimetric Determination of Heat Capacity, Entropy and Enthalpy of Mixed Oxides in the System
CaO–SrO–Bi$_2$O$_3$–Nb$_2$O$_5$–Ta$_2$O$_5$

177

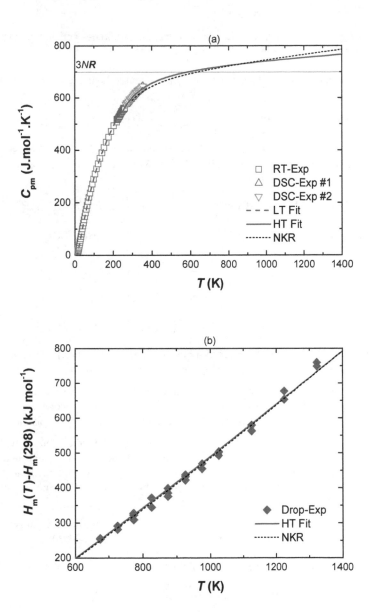

Figure 2. Temperature dependence of heat capacity (a) and relative enthalpy (b) of Bi$_7$Ta$_3$O$_{18}$ (3NR means the Dulong-Petit limit).

2.4. Evaluation of enthalpy of formation

The heats of drop-solution for the calcium and strontium carbonates and for the bismuth and niobium oxides were measured first. These data are necessary for the evaluation of the $\Delta_{ox}H$ values for the mixed oxides, and furthermore, these data could be compared with the literature data [49–52]. For the AECO₃ carbonates, the measured heat effect consists of three contributions:

$$\Delta_{ds}H\left(AECO_3,T\right)=\Delta_T H\left(AECO_3,T_0 \rightarrow T\right)+\Delta_{decomp}H\left(AECO_3,T\right)+\Delta_{sol}H\left(AEO,T\right) \quad (10)$$

The measurements were performed at 973 K. The values of $\Delta_{ds}H$(AECO₃, 973 K) are given in Table 6 along with the values of $\Delta_{ds}H$(AEO, 973 K), which were derived based on the following thermochemical cycle ($T_0 \approx 298$ K):

$$AECO_3(s,T_0) \quad \rightarrow \quad AEO(melt,T) + CO_2(g,T), \quad \Delta_{ds}H(AECO_3) \quad (11)$$

$$AECO_3(s,T_0) \quad \rightarrow \quad AEO(s,T_0) + CO_2(g,T_0), \quad \Delta_{decomp}H(AECO_3) \quad (12)$$

$$CO_2(g,T_0) \quad \rightarrow \quad CO_2(g,T), \quad \Delta_T H(CO_2) \quad (13)$$

$$AEO(s,T_0) \quad \rightarrow \quad AEO(melt,T), \quad \Delta_{ds}H(AEO) \quad (14)$$

$$\Delta_{ds}H(AEO) = \Delta_{ds}H(AECO_3) - \Delta_{decomp}H(AECO_3) - \Delta_T H(CO_2) \quad (15)$$

The values $\Delta_{decomp}H$(CaCO₃, 298 K) = 178.8 kJ mol⁻¹, $\Delta_{decomp}H$(SrCO₃, 298 K) = 233.9 kJ mol⁻¹ and $\Delta_T H$(CO₂, 298 → 973 K) = 32.0 kJ mol⁻¹ [53] were used for the calculations.

Next, the $\Delta_{ds}H$ values of the binary oxides Bi₂O₃ and Nb₂O₅ were measured. Because the dissolution of Nb₂O₅ and of the mixed oxides at 973 K proceeds rather slowly, the higher temperature of 1073 K was used. The measured values $\Delta_{ds}H$ are also given in Table 6.

The experimental values of $\Delta_{ds}H$ for SrCO₃ and CaCO₃ are in quite good agreement with the literature data [49–51]. On the other hand, our results and the published [52] values of $\Delta_{ds}H$(Nb₂O₅) are quite different. It should be noted that a more endothermic value $\Delta_{decomp}H$(SrCO₃, 298 K) = 249.4 kJ mol⁻¹ is presented in the literature [45], which results in more exothermic value for $\Delta_{ds}H$(SrO) by 15.5 kJ mol⁻¹.

$\Delta_{ds}H$ for the mixed oxides was measured at 1073 K. The following thermochemical cycle was used for the calculation of $\Delta_{ox}H$ for calcium and strontium niobates ($T_0 \approx 298$ K):

$$AE_xNb_2O_{5+x}(s,T_0) \quad \rightarrow \quad xAEO(melt,T) + Nb_2O_5(melt,T), \quad \Delta_{ds}H(AEO) \quad (16)$$

$$AEO(s,T_0) \quad \rightarrow \quad AEO(melt,T), \quad \Delta_{ds}H(AEO) \quad (17)$$

$$Nb_2O_5(s,T_0) \quad \rightarrow \quad Nb_2O_5(melt,T), \quad \Delta_{ds}H(Nb_2O_5) \quad (18)$$

$$xAEO(s,T_0) + Nb_2O_5(s,T_0) \quad \rightarrow \quad AE_xNb_2O_{5+x}(s,T_0), \quad \Delta_{ox}H(AE_xNb_2O_{5+x}) \quad (19)$$

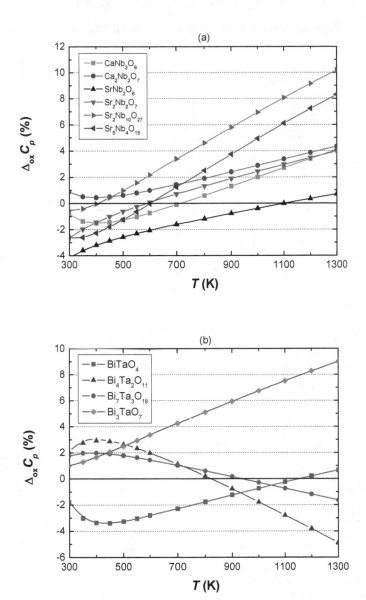

Figure 3. Temperature dependences of $\Delta_{ox}C_{pm}$ for various mixed oxides in the systems CaO–Nb₂O₅, and SrO–Nb₂O₅ (a) and Bi₂O₃–Ta₂O₅ (b)

$$\Delta_{ox}H(AE_xNb_2O_{5+x}) = x\Delta_{ds}H(AEO) + \Delta_{ds}H(Nb_2O_5) - \Delta_{ds}H(AE_xNb_2O_{5+x}) \qquad (20)$$

An analogous scheme was applied to calculate $\Delta_{ox}H(BiNbO_4)$. All of the experimental and calculated values are summarized in Table 7. The $\Delta_{ox}H(298\ K)$ values derived from high-temperature EMN measurements [7,8,11] for the CaO-Nb$_2$O$_5$ oxides and the assessed values from the phase diagram for the SrO-Nb$_2$O$_5$ oxides [12] are also presented in Table 7.

Substance	T (K)	$\Delta_{ds}H$ (kJ mol^{-1}) [a]	$\Delta_{ds}H$ (kJ mol^{-1})
CaCO$_3$	973	128.4 ± 10.1 (10)	119.70 ± 1.02 [b]
CaO	973	−82.39	−90.70 ± 1.69 [b]
CaO	1073	−77.04 [c]	
SrCO$_3$	973	131.4 ± 9.1 (7)	130.16 ± 1.66 [d] 134.48 ± 1.89 [e]
SrO	973	−134.47	−135.82 ± 2.48 [d] −131.42 ± 1.89 [e]
SrO	1073	−129.25 [f]	
Bi$_2$O$_3$	973	26.0 ± 2.9 (12)	
Bi$_2$O$_3$	1073	39.6 [g]	
Nb$_2$O$_5$	1073	141.8 ± 6.0 (11)	91.97 ± 0.78 [h]

[a] Data from the present work. The uncertainty is two standard deviations of the mean (95% confidence level), the number in parentheses is the number of experiments performed, [b] From ref. [49], $T = 976$ K, [c] The value $\Delta_TH(CaO, 973 \rightarrow 1073\ K) = 5.35$ kJ mol^{-1} [26] was used for the calculation, [d] From ref. [50], $T = 975$ K, [e] From ref. [51], $T = 974$ K, [f] The value $\Delta_TH(SrO, 973 \rightarrow 1073\ K) = 5.35$ kJ mol^{-1} [27] was used for the calculation, [g] The value $\Delta_TH(Bi_2O_3, 973 \rightarrow 1073\ K) = 13.61$ kJ mol^{-1} [28] was used for the calculation, [h] From ref. [52], $T = 973$ K.

Table 6. Enthalpy of drop-solution in 3Na$_2$O + 4MoO$_3$ melts [30]

Substance	T (K)	$\Delta_{ds}H$ (kJ mol^{-1}) [a]	$\Delta_{ox}H(298\ K)$ (kJ mol^{-1}) [b]	$\Delta_{ox}H$ (298 K) (kJ mol^{-1})
CaNb$_2$O$_6$	1073	196.8 ± 20.7 (8)	−132.0 ± 23.8	−159.8 [c] −130.1 [d]
Ca$_2$Nb$_2$O$_7$	1073	195.7 ± 27.8 (8)	−208.0 ± 31.9	−147.3 [c] −177.5 [e]
SrNb$_2$O$_6$	1073	180.50 ± 15.7 (4)	−167.9 ± 19.1	−325.0 [f]
Sr$_2$Nb$_2$O$_7$	1073	167.54 ± 34.7 (4)	−289.2 ± 37.5	−367.4 [f]
BiNbO$_4$	1073	132.61 ± 8.9 (7)	−41.9 ± 11.1	

[a] Data from the present work. The uncertainty is two standard deviations of the mean (95% confidence level), the number in parentheses is the number of experiments performed, [b] The experimental data from the present work. The uncertainty was calculated according to the error propagation law, [c] From ref. [11], [d] From ref. [7], [e] From ref. [8], [f] From ref. [12].

Table 7. Enthalpies of drop-solution in 3Na$_2$O + 4MoO$_3$ melt ($\Delta_{ds}H$) and enthalpy of formation from constituent binary oxides ($\Delta_{ox}H$) [30]

Our values for the calcium niobates are in good agreement with Raghavan's data [7,8], while the data from Dneprova et al. [11] are quite different. Moreover, a relation, $\Delta_{ox}H(CaNb_2O_6) > \Delta_{ox}H(Ca_2Nb_2O_7)$, that holds for the values from the work of Dneprova et al. is rather unexpected. The $\Delta_{ox}H$ values for strontium niobates obtained based on the binary SrO–Nb₂O₅ phase diagram evaluation [12] are substantially more exothermic than our calorimetric data. These large differences in the $\Delta_{ox}H$ values are not surprising in view of simultaneous differences in the $\Delta_{ox}S$ values from the assessment [12] and those derived from low temperature dependences of the molar heat capacity of SrNb₂O₆ and Sr₂Nb₂O₇ [24,26].

3. Empirical correlation S–V

A linear correlation between the standard molar entropy at 298.15 K and the formula unit volume $V_{f.u}$ has been proposed by Jenkins and Glaser [54–56]. This approach was used in this work for mixed oxides in the CaO–SrO–Bi₂O₃–Nb₂O₅–Ta₂O₅ system. The linear relation is obvious (see Fig. 4) and the straight line almost naturally passes through the origin:

$$S_m (J\,K^{-1} mol^{-1}) = 1680.5\, V_{f.u.} (nm^3\, f.u.^{-1}) \tag{21}$$

The average relative error in entropy is 8.2 %, the binary oxides CaO and Nb₂O₅ show the deviations around 20 %. It should be noted that, in this set of values, the simple analogy of NKR (Eq.(9)) provides a better prediction with an average relative error in entropy of 4.2 %.

Eq. (21) can be used for estimation of missing data. So, the estimated value $S_m(Sr_2Ta_2O_7)$ = 256.06 J K⁻¹ mol⁻¹ can be compared with the value 245.41 J K⁻¹ mol⁻¹ obtained by numeric integration of the $C_{pm}(T)/T$ dependences from zero to 298.15 K given in Ref. [1] (relative deviation of –4.3 %). Simple calculation $S_m(Sr_2Ta_2O_5) = 2S_m(SrO) + S_m(Ta_2O_5) = 250.25$ J K⁻¹ mol⁻¹ gives more reliable value (relative deviation 2.0 %).

4. Empirical estimation of enthalpy of formation

There are other mixed oxides in the system CaO–SrO–Bi₂O₃–Nb₂O₅–Ta₂O₅ for which the values of enthalpy of formation $\Delta_f H$ or enthalpy of formation from binary oxides $\Delta_{ox}H$ have not yet been determined. As a rough estimate, the values of $\Delta_{ox}H$ calculated according to an empirical method proposed by the authors [56] can be used. In the case of Ca, Sr and Bi niobates the following relation holds for $\Delta_{ox}H$:

$$\frac{\Delta_{ox}H}{n_{Nb} + n_{Me}} = -2 \cdot 96.5\, \alpha y\, x_{Nb}\, x_{Me}^{\delta}\, (X_{Nb} - X_{Me})^2 \tag{22}$$

where X_{Nb} and X_{Me} (Me = Ca, Sr or Bi) are Pauling's electronegativities of the relevant elements, x_{Nb} and x_{Me} are the molar fractions of the oxide-forming elements ($x_{Nb} = n_{Nb}/(n_{Nb} + n_{Me})$ etc.), y is the number of oxygen atoms per one atom of oxide-forming elements and α

and δ are the model parameters. Using Pauling's electronegativities, $X_{Nb} = 1.60$, $X_{Ca} = 1.00$, $X_{Sr} = 0.95$, and $X_{Bi} = 2.02$, and the calorimetric values of $\Delta_{ox}H$ obtained in this work, the values of $\alpha = 2.576$ and $\delta = 1.50$ were derived from the least-squares fit. The estimated $\Delta_{ox}H$ values for calcium and strontium niobates are shown in Fig. 5. The values of $\Delta_{ox}H$ that were calculated according to an empirical method proposed by Zhuang et al. [57] are displayed for comparison.

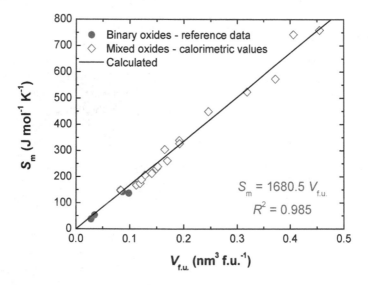

Figure 4. Correlation between the standard molar entropy at 298.15 K and the formula unit volume $V_{f.u.}$ for various mixed oxides in the system CaO–SrO– Bi_2O_3–Nb_2O_5–Ta_2O_5 (data from table Table 3)

5. Conclusion

The above presented data derived from calorimetric measurements became the basis for thermodynamic database FS-FEROX [58] compatible with the FactSage software [59,60]. Missing data for other stoichiometric mixed oxides were estimated by the empirical methods described before: the Neumann-Kopp's rule for heat capacities, the entropy-volume correlation for molar entropies and electronegativity-differences method for enthalpies of formation. At the same time, thermodynamic description of a multicomponent oxide melt was obtained analyzing relevant binary phase diagrams published in literature. The database and the FactSage software were subsequently used for various equilibrium calculations including binary T-x phase diagrams and ternary phase diagrams in subsolidus region. Thermodynamic modeling of $SrBi_2Ta_2O_9$ and $SrBi_2Nb_2O_9$ thin layers deposition from the gaseous phase were also performed to optimize the deposition conditions.

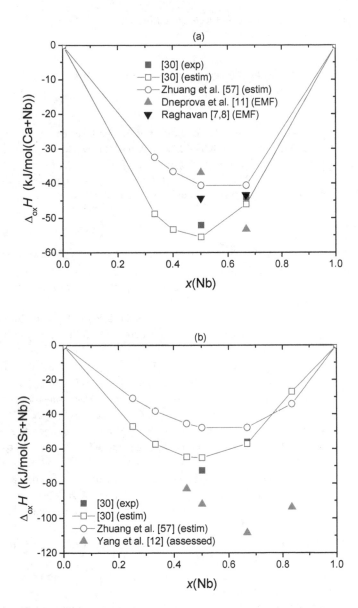

Figure 5. Values of enthalpy of formation of the mixed oxides from constituent binary oxides in the CaO–Nb$_2$O$_5$ (a) and SrO–Nb$_2$O$_5$ (b) systems (lines serve only as a guide for the eyes)

Author details

Jindřich Leitner, David Sedmidubský and Květoslav Růžička
Institute of Chemical Technology, Prague, Czech Republic

Pavel Svoboda
Charles University in Prague, Faculty of Mathematics and Physics, Prague, Czech Republic

Acknowledgment

This work was supported by the Ministry of Education of the Czech Republic (research projects N° MSM6046137302 and N° MSM6046137307). Part of this work was also supported from the Grant Agency of the Czech Republic, grant No P108/10/1006. Low temperature experiments were performed in MLTL (http://mltl.eu/), which is supported within the program of Czech Research Infrastructures (project no. LM2011025).

6. References

[1]   Akishige Y, Shigematsu H, Tojo T, Kawaji H, Atake T (2005) Specific heat of $Sr_2Nb_2O_7$ and $Sr_2Ta_2O_7$. J. Therm. Anal. Calorim. 81: 537–540.

[2]   Shabbir G, Kojima S (2003) Acoustic and thermal properties of strontium pyroniobate single crystals. J. Phys. D: Appl. Phys. 36: 1036–1039.

[3]   Onodera A, Yoshio K, Myint CC, Kojima S, Yamashita H, Takama T (1999) Thermal and structural studies of phase transitions in layered perovskite $SrBi_2Ta_2O_9$. Jpn. J. Appl. Phys. 38: 5683–5685.

[4]   Onodera A, Yoshio K, Myint CC, Tanaka M, Hironaka K, Kojima S (2000) Thermal behavior in feroelectric $SrBi_2Ta_2O_9$ thin films. Ferroelectrics 241: 159–166.

[5]   Yoshio K, Onodera A, Yamashita H (2003) Ferroelectric phase transition and new intermediate phase in bi-layered perovskite $SrBi_2Ta_2O_9$. Ferroelectrics 284: 65–74.

[6]   Morimoto K, Sawai S, Hisano K, Yamamoto T (1999) Simultaneous measurements of specific heat capacity and dielectric constant of ferroelectric $SrBi_2(Nb_xTa_{1-x})_2O_9$ ceramics. Ferroelectrics 227: 133–140.

[7]   Raghavan S (1991) Thermodynamic stability of monocalcium niobate using calcium fluoride solid electrolyte galvanic cell. Trans. Indian Inst. Met. 44: 285–286.

[8]   Raghavan S (1992) Thermodynamics of formation of high calcium niobates from emf measurements. J. Alloys Compd. 179: L25–L27.

[9]   Raghavan S (1991) Electrochemical determination of the stability of calcium ditantalate. Indian J. Technol. 29: 313–314.

[10] Raghavan S (1992) Thermodynamics of the formation of high calcium tantalates from emf measurements. J. Alloys Compd. 189: L39–L40.

[11] Dneprova VG, Rezukhina TN, Gerasimov YI (1968) Thermodynamic properties of some calcium niobates. Dokl. Akad. Nauk SSSR 178: 135–137.

[12] Yang Y, Yu H, Jin Z (1999) Thermodynamic calculation of the SrO-Nb$_2$O$_5$ system.J. Mater. Sci. Technol. 15: 203– 207.

[13] Saunders N, Miodownik AP (1998) Calphad (Calculation of phase diagrams): A comprehensive guide. Pergamon Materials Series, Vol. 1., Pergamon.

[14] Hallstedt B, Risold D, Gauckler LJ (1997): Thermodynamic assessment of the bismuth–calcium–oxygen system. J. Am. Ceram. Soc. 80: 2629–2636.

[15] Hallstedt B, Risold D, Gauckler LJ (1997) Thermodynamic assessment of the bismuth–strontium–oxygen system. J. Am. Ceram. Soc. 80: 1085–1094.

[16] Idemoto Y, Shizuka K, Yasuda Y, Fueki K (1993) Standard enthalpies of formation of member oxides in the Bi–Sr–Ca–Cu–O system. Physica C 211: 36–44.

[17] Jacob KT, Jayadevan KP (1997) Combined use of oxide and fluoride solid electrolytes for the measurement of Gibbs energy of formation of ternary oxides: System Bi–Ca–O. Mater. Trans. JIM 38: 427–436.

[18] Jacob KT, Jayadevan KP (198) System Bi–Sr–O: Synergistic measurements of thermodynamic properties using oxide and fluoride solid electrolytes. J. Mater. Res. 13: 1905–1908.

[19] Hallstedt B, Gauckler L (2003) Revision of the thermodynamic descriptions of the Cu–O, Ag–O, Ag–Cu–O, Bi–Sr–O, Bi–Ca–O, Bi–Cu–O, Sr–Cu–O, Ca–Cu–O and Sr–Ca–Cu–O systems CALPHAD 27: 177–191.

[20] Abrman P, Sedmidubský D, Strejc A, Voňka P, Leitner J (2002) Heat capacity of mixed oxides in the Bi$_2$O$_3$–CaO system. Thermochim. Acta 381: 1–7.

[21] Hampl M, Strejc A, Sedmidubský D, Růžička K, Hejtmánek J, Leitner J (2006) Heat capacity, enthalpy and entropy of bismuth niobate and bismuth tantalate. J. Solid State Chem. 179: 77–80.

[22] Leitner J, ,Hampl M, Růžička K, Sedmidubský D, Svoboda P, Vejpravová J (2006) Heat capacity, enthalpy and entropy of strontium bismuth niobate and strontium bismuth tantalate. Thermochim. Acta 450: 105–109.

[23] Hampl M, Leitner J, Růžička K, Straka M, Svoboda P (2007) Heat capacity and heat content of BiNb$_5$O$_{14}$. J. Thermal Anal. Calorimetry 87: 553–556.

[24] Leitner J, Hampl M, Růžička K, Straka M, Sedmidubský D, Svoboda P. (2008) Thermodynamic properties of strontium metaniobate SrNb$_2$O$_6$. J. Thermal. Anal. Calorimetry 91: 985–990.

[25] Leitner J, Hampl M, Růžička K, Straka M, Sedmidubský D, Svoboda P (2008) Heat capacity, enthalpy and entropy of strontium niobate Sr$_2$Nb$_2$O$_7$ and calcium niobate Ca$_2$Nb$_2$O$_7$. Thermochim. Acta 475: 33–38.

[26] Leitner J, Růžička K, Sedmidubský D, Svoboda P (2009) Heat capacity, enthalpy and entropy of calcium niobates. J. Thermal. Anal. Calorimetry 95: 397–402.

[27] Leitner J, Šipula I, Růžička K, Sedmidubský D, Svoboda P (2009) Heat capacity, enthalpy and entropy of strontium niobates Sr$_2$Nb$_{10}$O$_{27}$ and Sr$_5$Nb$_4$O$_{15}$. J. Alloys Compd. 481: 35–39.

[28] Leitner J, Jakeš V, Sofer Z, Sedmidubský D, Růžička K, Svoboda P (2011) Heat capacity, enthalpy and entropy of ternary bismuth tantalum oxides. J. Solid State Chem. 184: 241–245.

[29] Leitner J, Sedmidubský D, Růžička K, Svoboda P (2012) Heat capacity, enthalpy and entropy of $SrBi_2O_4$ and $Sr_2Bi_2O_5$. Thermochim. Acta. 531: 60–65.

[30] Leitner J, Nevřiva M, Sedmidubský D, Voňka P (2011) Enthalpy of formation of selected mixed oxides in a $CaO–SrO–Bi_2O_3–Nb_2O_5$ system. J. Alloys Compd. 509: 4940–4943.

[31] Lashley J.C., Hundley MF, Migliori A, Sarrao JL, Pagliuso PG, Darling TW, Jaime M, Cooley JC, Hults WL, Morales L, Thoma DJ, Smith JL, Boerio-Goates J, Woodward BF, Stewart GR, Fisher RA, Phillips NE (2003) Critical examination of heat capacity measurements made on a Quantum Design physical property measurement system. Cryogenics 43: 369–378.

[32] Dachs E, Bertoldi C (2005) Precision and accuracy of the heat-pulse calorimetric technique: low-temperature heat capacities of milligram-sized synthetic mineral samples. Eur. J. Mineral. 17: 251–259.

[33] Marriott RA, Stancescu M, Kennedy CA, White MA (2006) Technique for determination of accurate heat capacities of volatile, powdered, or air-sensitive samples using relaxation calorimetry. Rev. Sci. Instrum. 77: 096108 (3 pp).

[34] Kennedy CA, Stancescu M, Marriott RA, White MA (2007) Recommendations for accurate heat capacity measurements using a Quantum Design physical property measurement system. Cryogenics 47: 107–112.

[35] Shi Q, Snow CL, Boerio-Goates J, Woodfield BF (2010) Accurate heat capacity measurements on powdered samples using a Quantum Design physical property measurement system. J. Chem. Thermodyn. 42: 1107–1115.

[36] Hwang JS, Lin KJ, Tien C (1997) Measurement of heat capacity by fitting the whole temperature response of a heat-pulse calorimeter. Rev. Sci. Instrum. 68: 94–101.

[37] Schnelle W, Engelhardt J, Gmelin E (1999) Specific heat capacity of Apiezon N high vacuum grease and of Duran borosilicate glass. Cryogenics 39: 271–275.

[38] Quantum Design, Physical Property Measurement System – Application Note, http://www.qdusa.com/sitedocs/productBrochures/heatcapacity-he3.pdf (accessed 21-12-2011).

[39] Arblaster JW (1994) The thermodynamic properties of platinum on ITS-90, Platinum Metals Rev. 38: 119–125.

[40] Rodriguez-Carvajal J (1993) Recent advances in magnetic structure determination by neutron powder diffraction. Physica B 192: 55–69.

[41] Martin CA (1991) Simple treatment of anharmonic effects on the specific heat. J. Phys.: Condens. Matter 3: 5967–5974.

[42] Taylor JR, Dinsdale AT (1990) Thermodynamic and phase diagram data for the $CaO–SiO_2$ system. CALPHAD 14: 71–88.

[43] Risold D, Hallstedt B, Gauckler LJ (1996) The strontium-oxygen system. CALPHAD 20: 353–361.

[44] Risold D, Hallstedt B, Gauckler LJ, Lukas HL, Fries SG (1995) The bismuth-oxygen system. J. Phase Equilib. 16: 223–234.

[45] Knacke O, Kubaschewski O, Hesselmann K (1991) Thermochemical Properties of Inorganic Substances, 2nd Ed. Berlin: Springer.

[46] Qiu L, White MA (2001) The constituent additivity method to estimate heat capacities of complex inorganic solids. J. Chem. Educ. 78: 1076–1079.

[47] Leitner J, Chuchvalec P, Sedmidubský D, Strejc A, Abrman P (2003) Estimation of heat capacities of solid mixed oxides. Thermochim. Acta 395: 27–46.

[48] Leitner J, Voňka P, Sedmidubský D, Svoboda P (2010) Application of Neumann-Kopp rule for the estimation of heat capacity of mixed oxides. Thermochim. Acta 497: 7–13.

[49] Helean KB, Navrotsky A, Vance ER, Carter ML, Ebbinghaus B, Krikorian O, Lian J, Wang LM, Catalano JG (2002) Enthalpies of formation of Ce-pyrochlore, $Ca_{0.93}Ce_{1.00}Ti_{2.035}O_{7.00}$, U-pyrochlore, $Ca_{1.46}U^{4+}_{0.23}U^{6+}_{0.46}Ti_{1.85}O_{7.00}$ and Gd-pyrochlore, $Gd_2Ti_2O_7$: three materials relevant to the proposed waste form for excess weapons plutonium. J. Nuclear Mater. 303: 226–239.

[50] Cheng J, Navrotsky A (2004) Energetics of magnesium, strontium, and barium doped lanthanum gallate perovskites. J. Solid State Chem. 177: 126–133.

[51] Xu H, Navrotsky A, Su Y, Balmer ML (2005) Perovskite Solid Solutions along the NaNbO3-SrTiO3 Join: Phase Transitions, Formation Enthalpies, and Implications for General Perovskite Energetics. Chem. Mater. 17: 1880–1886.

[52] Pozdnyakova I, Navrotsky A, Shilkina L, Reznitchenko L (2002) Thermodynamic and structural properties of sodium lithium niobate solid solutions. J. Am. Ceram. Soc. 85: 379–384.

[53] Robbie RA, Hemingway BS (1995) Thermodynamic properties of minerals and related substances at 298.15 K and 1 bar pressure and at higher temperatures, U.S. Geological Survey Bulletin, Vol. 2131,Washington.

[54] Jenkins HDB, Glasser L (2003) Standard absolute entropy, $S°_{298}$, values from volume or density. 1. Inorganic materials. Inorg. Chem. 42: 8702–8708.

[55] Jenkins HDB, Glasser L (2006) Volume-based thermodynamics: Estimations for 2:2 salts, Inorg. Chem. 45: 1754–1756.

[56] Voňka P, Leitner J (2009) A method for the estimation of the enthalpy of formation of mixed oxides in Al₂O₃–Ln₂O₃ systems. J. Solid State Chem. 182: 744–748.

[57] Zhuang W, Liang J, Qiao Z, Shen J, Shi Y, Rao G (1998) Estimation of the standard enthalpy of formation of double oxide. , J. Alloys Compd. 267: 6-10.

[58] Leitner J, Sedmidubský D, Voňka P. (2009) Thermodynamic database for the oxide system CaO–SrO–Bi₂O₃–Nb₂O₅–Ta₂O₅. CALPHAD XXXVIII, 17.-22.5.2009, Praha, ČR.

[59] Bale CW, Chartrand P, Degterov SA, Eriksson G, Hack K, Ben Mahfoud R, Melançon J, Pelton AD, Petersen S (2002) FactSage thermochemical software and databases. Calphad 26: 189-228.

[60] Bale CW, Bélisle E, Chartrand P, Decterov SA, Eriksson G, Hack K, Jung IH, Kang YB, Melançon J, Pelton AD, Robelin C, Petersen S (2009) FactSage thermochemical software and databases — recent developments. Calphad 33: 295-311.

Chapter 9

Oxidative Stability of Fats and Oils Measured by Differential Scanning Calorimetry for Food and Industrial Applications

M.D.A. Saldaña and S.I. Martínez-Monteagudo

Additional information is available at the end of the chapter

1. Introduction

Fats and oils are important ingredients in the human diet for nutritional and sensory contributions. The terms fats and oils commonly refer to their phase being solid and liquid, respectively. In addition, lipids are the main ingredients to manufacture various products, such as soups, butter, ready to eat food, among others for the food industry and other products, such as lipstick, creams, etc, for the cosmetic and pharmaceutical industries. Most of these products use vegetable oils from seeds, beans, and nuts, which are important due to their high content in polyunsaturated fatty acids compared to animal fats. However, oxidation of unsaturated fatty acids is the main reaction responsible for the lipid degradation, which is related to the final quality of the product. Furthermore, lipids undergo oxidation, developing unpleasant taste, off flavour and undesirable changes in quality, decreasing the nutritional value of the product and compromising safety of the product that might even affect health and well-being.

In general, oxygen reacts with the double bonds present in lipids, following a free radical mechanism, known as autooxidation. This reaction is quite complex and depends on the lipid type used and the processing conditions. The use of thermal processes, such as frying, sterilization, hydrolysis, etc, accelerate the oxidation of lipids. Various different reactions during lipid oxidation occur simultaneously at different rates. These reactions release heat that can be measured using differential scanning calorimetry (DSC).

Oxidation temperatures and kinetic parameters obtained from DSC can be used to rank and classify lipids in terms of their oxidative stability. Therefore, the reproducibility of oxidation experiments is crucial to evaluate the oxidative stability of lipids using DSC since variables, such as pre-treatment and amount of sample, the heating protocol, among others, strongly influence the results. Other methods that assess the extent of oxidative deterioration are peroxide value (PV) that measures volumetrically the concentration of hydroperoxides, anisidine value (AV), spectrophotometric measurements in the UV region and gas chromatography (GC) analysis for volatile compounds [1-3]. Over the years, thousands of studies have focused on monitoring and evaluating the oxidation of lipids using the Rancimat method, PV, AV, spectrophotometric and GC analysis of fats and oils from various sources. However, it is beyond the scope of this chapter to provide a comprehensive listing of all research using those methods. Comprehensive reviews on the oxidative lipid deterioration using those methods are well discussed somewhere else [4-6].

Among all these methods that measure the extent of lipid oxidation, DSC is widely used as an analytical, diagnostic and research tool from which relevant information, such as onset temperature of oxidation (T_{on}), height, shape and position of peaks are obtained and used for subsequent kinetic calculations. Kinetic information of lipid oxidation has been reported for a number of lipid systems, such as soybean/anhydrous milk fat blends, unsaturated fatty acids (oleic, linoleic, and linolenic acids), saturated fatty acids (lauric, myristic, palmitic, and stearic acids), "natural" vegetable oils (canola, corn, cottonseed, and soybean oils) and genetically modified vegetable oils.

This chapter focuses on the principles of lipid oxidation, the use of DSC technique to evaluate lipid oxidation, and recent studies on oxidative stability of fats and oils for food and industrial applications, addressing the generation and analysis of DSC thermograms for kinetic studies, where a method to analyse DSC data is described in detail, as well as the interpretation of kinetic parameters obtained at isothermal and non-isothermal conditions. In addition, some results on oxidation kinetics of milk fat after the use of traditional and emerging technologies, such as enzymatic hydrolysis and pressure assisted thermal processing are discussed in detail. Finally, conclusions on lipid oxidation analysis by DSC are provided.

2. Fundamentals of lipid oxidation

Lipid oxidation is a free radical chain reaction that leads to the development of unpleasant flavour and taste, loss of nutrients and formation of toxic compounds [7, 8]. Consequently, the shelf life and the final use of any lipid depend on its resistance to oxidation or oxidative stability [9]. The term lipid oxidation usually refers to a three consecutive reactions or stages, known as initiation, propagation and termination (Figure 1). In the initiation stage, free radicals are formed through thermolysis, where the break of covalent bonds is induced by heat. In addition, free radicals can also be formed due to the presence of enzymes, light, metal ions (Ca^{2+} and Fe^{3+}) and reactive oxygen species. A list of initiators of lipid oxidation and their standard reduction potentials are provided somewhere else [10]. Compounds that homolyze at relative low temperature (<100°C) are important initiators of radical-based chain reactions. Unsaturated fatty acids are compounds that homolyze at lower temperatures compare to saturated fatty acids. The homolytic products of unsaturated fatty acids are hydroxyl radical (HO•), alkyl radical (RO•) and hydroperoxyl radical (HOO•), which further reacts with triplet oxygen to form peroxyl radicals. Among the starters, hydroxyl radicals are mainly responsible for the initiation of lipid oxidation due to its strong tendency to acquire electrons [10]. These radical products that have high energy, bond to a hydrogen molecule from the lipid structure, forming hydroperoxides (primary oxidation products). The formation of hydroperoxides can be repeated several times, propagating the oxidation reactions. Conjugated fatty acids have more than one type of primary oxidation products and more than one oxidation pathway, as previously reported [11, 12]. A kinetic analysis on autoxidation of methyl-conjugated linoleate showed that monomeric and cyclic peroxides are the major primary oxidation products rather than hydroperoxides [12]. Consequently, addition by Diels Alder-type reaction was earlier suggested as a reaction mechanism.

The next oxidation stage is the propagation, which consists in the further degradation of hydroperoxides or any other primary oxidation product [1]. There are mainly two types of degradation products from hydroperoxides. First, hydroperoxides interact with double bonds to form monomeric degradation products, such as ketones. The reaction occurs through the reduction of the hydroperoxyl group to hydroxyl derivative. Second, low molecular weight products, that results from the cleavage of the hydroperoxide chain, form aldehydes, ketones, alcohols and hydrocarbons. These low molecular weight compounds are responsible for the rancid and off-flavour produced by oxidized fats [1-4]. Finally, hydroperoxides and primary oxidation products homolyze to form peroxyl or alkoxy radicals that further reacts to form stable dimer-like products. In addition, alcohols, and unsaturated fatty acids (secondary oxidation products) also lead to termination products. The resulting compounds form viscous materials through polymerization as the oxidation proceeds. These polymers are oil insoluble and represent the termination stage of oxidation [2, 15].

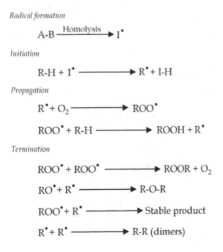

Figure 1. Main lipid oxidation reactions.

3. Fundamentals on the use of DSC to study lipid oxidation

The oxidation mechanisms presented in Figure 1 is an oversimplification because lipids consist of a non-homogeneous mixture of fatty acids. For example, anhydrous milk fat (AMF) is composed of more than 400 fatty acids with extremely diverse chain lengths, position and number of unsaturations of their fatty acids [13, 14]. Consequently, several reactions occur simultaneously at different rates as the oxidation proceeds. Although several methods have been used to analyze and monitor lipid oxidation [1], the oxidation reactions cannot be measured by a single method due to their complexity. Some of the available methods allow quantifying one or more reaction products of the different oxidation stages. Methods, such as oxidative stability index (OSI) and peroxide value (PV) are officially accepted by the American of Analytical Communities (AOAC) [1-6], while other methods are routinely used, such as the Racimat, chemilominescent, and volumetric methods [16].

An important overlooked characteristic of the oxidation reactions of lipids is the exothermal effect as the oxidation occurs. The released heat from a particular reaction can be measured using DSC in either isothermal or non-isothermal mode. For DSC oxidation measurements, the heat released from the oxidized oil is compared to the heat flowing from an inert reference (empty pan) both heated at the same rate. When the oxidation of the sample occurs, the recorded heat shows a peak which area is proportional to the amount of heat released by the sample. Figure 2 shows an ideal thermogram for the non-isothermal oil oxidation with the three consecutive reaction stages of initiation, propagation and termination.

The heat released by the oxidized oil is recorded as the heat flow signal (y-axis) as a function of temperature (x-axis). The period of time where no change in the heat flow signal occurs is known as the induction time (Figure 2) that is exemplified at the beginning of the thermogram. An excellent review on the theory and application of induction time is provided elsewhere [17]. The length of the induction time is often considered as a measurement of oil stability. During this period, no chemical reaction occurs. At the point in

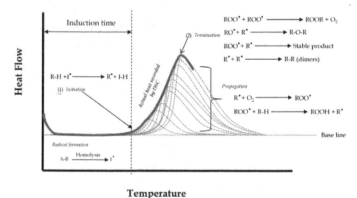

Figure 2. Ideal thermogram of non-isothermal oil oxidation.

which the heat flow signal separates from the baseline (straight line) is considered to be the end of the induction time (arrow (1)). Arrow (1) also indicates the start of oxidation or initiation stage. This stage is short and can be theoretically interpreted as the reaction between the radical, formed during the induction time, and the unsaturated fatty acid. The products of this reaction are unstable hydroperoxides that further react propagating the oxidation. A sudden increase in the heat flow signal is related to the propagation stage. The blue dashed lines illustrate oxidation reactions that occur and cannot be detected by the DSC because they are less exothermal. Finally, arrow (2) illustrates the termination stage, where stable products are formed. The red line is the actual heat flow recorded by the DSC.

In the next section, a set of recommended guidelines and laboratory practices are reviewed for the good use and analysis of data using DSC.

4. Important considerations for measurement of lipid oxidation using DSC

Although DSC is a simple, convenient and fast technique to measure lipid oxidation, some recommended guidelines should be considered to obtain reliable and reproducible results. Among those guidelines are the pre-treatment and sample preparation, amount of sample used, heating protocol, gas flow rate, and interpretation of DSC thermograms.

Sample pre-treatment – a representative amount of lipid sample should be used for DSC oxidation measurements. The lipid should be melted at a temperature that ensures that its thermal memory is erased. The melting temperature for lipids is quite diverse, ranging from -25 to 80°C. Melting temperatures of some vegetable oils are provided elsewhere [18]. In some lipid-based products, the pre-treatment of sample involves the fat extraction from the food matrix. For example, fat from commercial baby formulas was first extracted with chloroform [19], and then the fat was further dried under vacuum. Therefore, the behavior of these extracted fats might be different from the fat in the original matrix. In addition, a proper chemical description of any pre-treatment must be provided together with the thermal history.

Sample preparation – a liquid sample should be loaded into the DSC pan using a syringe or a Pasteur pipette. The lipid oxidation measured by DSC can be conducted in an open aluminum pan or in a hermetic sealed pan with a pinhole (Figure 3). The main difference between an open pan and a sealed pan with a pinhole is the diffusion of oxygen and the amount of oxygen that is in contact with the sample. This is because the thermal conductivity of the air is smaller than that of the metal of the pan. Indeed, numerical simulations showed that the energy transmitted to the sample comes from the plate, which transmits the heat to the pan [20].

A comparison of glycerol oxidation obtained in an open pan and a hermetic sealed pan with a pinhole was earlier reported [21]. For experiments conducted in an open pan, the maximum heat flow temperature of oxidation of glycerol was around 40°C lower than those obtained with sealed pans. In an open pan, the vapor produced during lipid oxidation

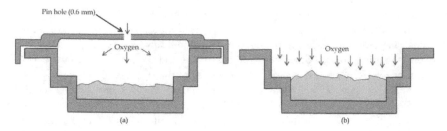

Figure 3. Illustrations of common DSC pans: (a) hermetically sealed pan with a pinhole, and (b) an open aluminium pan.

leaves the pan as it is formed. This is because the purge of oxygen acts as a carrier of the vapor. Consequently, part of the mass is lost before reaching the temperature at which the oxidation starts. On the other hand, in a sealed pan with a pinhole, the vapor produced cannot escape from the pan, remaining in the oil. This vapor increases the pressure inside the pan and elevates the oxidation temperature. Contradictory results were reported for the melting temperature of benzoic acid and vanillin obtained in an open pan and a sealed pan with a pinhole [22]. No significant differences were observed in the onset temperature (T_{on}) and the peak maximum temperature (T_P). Unfortunately, studies with a direct comparison between the types of pans (open and sealed with a pinhole) for oxidation of are scarce in the literature. Indeed, the international organization for standardization does not specify the types of pans for the determination of oxidation induction time [23]. Although hermetic sealed pans with a pinhole have an additional cost to each experimental run, their use avoids contamination of the DSC chamber.

Sample size – the amount of sample has significant effect on the shape of the thermogram and reproducibility of the DSC oxidation experiments as it is related to heat transfer within the pan. Figure 4 illustrates the effect of the amount of sample in hermetic sealed pans with a pinhole. For the sample with an optimum sample thickness (Figure 4a), the oil is in contact with excess of oxygen, facilitating oxygen diffusion within the oil sample. An earlier study [7] recommended 1 mm of sample thickness (approximately 1.5 mg of oil) to yield consistent results in non-isothermal oil oxidation. Similarly, no changes on the DSC thermograms using samples between 1 to 4 mg were reported [24]. A ratio of 1:3 (oil:oxygen) not only avoids diffusional limitations [6, 7, 25] but also allows the vapor molecules formed during the oxidation reaction to rapidly escape from the pan [26]. This enhances the baseline and the resolution of the oxidation thermogram [27].

On the other hand, an excess of sample (Figure 4b) creates a temperature gradient within the sample, especially at high heating rates [27]. Also, the diffusion of oxygen is limited, which broader the DSC oxidation curves. This is illustrated in Figure 5 where samples of 3 and 13 mg of anhydrous milk fat (AMF) were oxidized at 12°C/min from 100 to 250°C. Interestingly, the T_{on} of oxidation is quite similar between samples (181.12 and 179.33°C, respectively). But, the use of 13 mg of AMF leads to a broader and less resolved curve. The deviation of the heat flow signal from the vertical edge (broadening) is attributed to a longer time needed to oxidize a larger sample and the development of a temperature gradient within the sample. Another reason is that the probability of vapor molecules to escape from the pan through the pinhole is considerable reduced. Consequently, the peak maximum temperature shifts to a lower temperature because some of the vapor molecules react, accelerating the termination stage of oxidation.

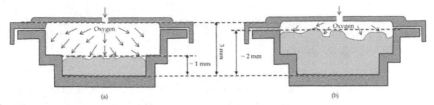

Figure 4. Illustration of (a) optimum sample size, and (b) excess sample size, in hermetic sealed pans with a pinhole.

Heating rate – this is one of the most important parameters to determine the oxidative stability and oxidation kinetics of lipids. Before the start of the heating rate, an equilibrium period between 3 and 5 min is recommended to enhance the baseline. At slow heating rates, primary oxidation products, such as hydroperoxides generated during the initial oxidation stage, react with excess of oxygen to form low molecular weight compounds (intermediate oxidation products), accelerating the degradation process. At fast heating rates, these intermediate products are lost through evaporation before they react with the lipid, shifting to a high value the threshold DSC signal [6, 9]. This phenomenon is the basis to calculate the kinetic parameters from a DSC thermogram [28]. However, it is important to highlight that the heating rate should not exceed 25°C/min since the temperature of the sample is different from the furnace temperature, creating a temperature gradient which affects the oxidation kinetics [29]. An important overlooked consideration is the temperature range at which the oxidation study should be conducted. In general, the temperature range should start and end as far as possible from the T_{on} and T_p (approximately a difference of at least 50°C).

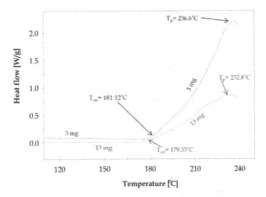

Figure 5. Effect of sample size on non-isothermal oxidation of anhydrous milk fat at 12°C/min. T_{on} – onset temperature of oxidation; T_p – maximum heat flow temperature.

DSC mode – oxidation experiments can be conducted in either isothermal or non-isothermal mode. Both methods provide analytical information, such as the oxidation induction time in the case of isothermal measurements and the oxidation onset temperature in the case of non-isothermal measurements [28, 29]. Further comparisons on the kinetic studies conducted in either isothermal or non-isothermal mode are discussed in the following sections.

5. Analysis of DSC thermograms

5.1. Location of key parameters

The analysis and interpretation of the generated DSC spectra consist in identifying key parameters, such as the induction time, onset and peak maximum temperatures. These key parameters are manually obtained from the DSC spectra. The T_{on} is obtained extrapolating the tangent drawn on the steepest slope of T_P. This procedure is usually performed

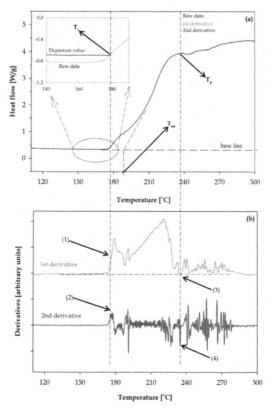

Figure 6. Anhydrous milk fat rich in conjugated linoleic acid oxidized at 15°C min[-1]. (a) Determination of the start temperature (T_s) (inlet), onset temperature (T_{on}) and maximum heat flow temperature (T_P), and (b) zoom on the first and second derivatives that precisely locates the T_s, T_{on}, and T_P

manually by the DSC operator, relying in the equipment software. Inherently, there is certain degree of uncertainty associated with this procedure. A method that accurately and unambiguously determines those key parameters from the DSC spectra was early proposed for lipid crystallization [30]. This method was first developed to calculate onset, offset, and peak maximum temperatures in crystallization of binary mixtures of different triacylglycerols. More recently, the same methodology was adapted to calculate the start, onset and peak maximum temperatures in non-isothermal oxidation of anhydrous milk fat [9]. Figure 6 exemplifies the location of the start, onset and peak maximum temperatures for anhydrous milk fat rich in conjugated linoleic acid oxidized at 15°C/min in a hermetic sealed pan with a pinhole.

Once the DSC curves are generated, the error associated with the raw data is calculated through standard deviation. Then, the first and second derivatives are calculated. The error is obtained from the baseline, which in Figure 6 corresponds to the segment of 140 to 178°C. This is essential since the signal variability can be misinterpreted as a thermal event. In this method, a true thermal event was considered when the heat flow signal is twice greater than the standard deviation of the baseline. This criterion is known as the departure value (inlet Figure 6a). To locate the start temperature of oxidation, three criteria were considered. Firstly, the first derivative of the signal shows an inflexion point between a maximum and a minimum point of the signal (arrow (1)). Secondly, the second derivative reaches a maximum point on the heat flow signal (arrow (2)). Finally, the heat flow signal should be greater than the departure value (inlet). T_P was obtained when the first derivative of the signal intersects with the x-axis (arrow (3)) and the second derivative reached a maximum point on the signal (arrow (4)). T_{on} was obtained extrapolating the tangent drawn on the steepest slope of T_P.

5.2. Iso-conversional method

In chemical reactions, the degree of conversion ($0 \leq \alpha \leq 1$) or extent of reaction of a particular compound is defined by moles at a given time divided by the initial moles [29]. Similarly, α in thermal analysis is defined by the heat flow at a given time or temperature divided by the heat flow at time or temperature at which the maximum heat flow signal is reached. The heat flow from the DSC spectra is converted to α based on the initial (signal$_o$) and final (signal$_f$) heat flow signals, as shown in equation (1).

$$\alpha = \frac{signal_o - signal}{signal_o - signal_f} \tag{1}$$

At a given degree of conversion, the reaction kinetics is described by a single-step reaction that follows an Arrhenius type equation within a narrow range of temperature. Then, the overall kinetics [31-34] is the result of multiple single-step reactions in the form:

$$\frac{d\alpha}{dt} = A \, exp^{\left(\frac{-E_a}{RT}\right)} f(\alpha) \tag{2}$$

where t is the time, A is the pre-exponential factor, E_a is the effective activation energy, T is the temperature, and $f(\alpha)$ is the reaction model. This procedure, known as iso-conversional or model-free method, is used to calculate kinetic triplet parameters (effective activation energy, E_a, pre-exponential factor, A and constant rate, k) in thermally stimulated reactions [29-34]. As known, reactions are the sequence of physical changes that can be measured by thermal techniques.

In non-isothermal oxidation of lipids, the consumption of oxygen can be neglected due to the large excess of oxygen generated by a constant flow rate (>25 mL/min). Such condition allows the formation of peroxides, being independently of the oxygen concentration, which also means that the autoxidation is a first order reaction [7, 9]. This is an essential assumption for the calculation of the kinetic triplet parameters (E_a, A, and k). A commonly used iso-conversional method is the Ozawa-Flynn-Wall method. Using this method, a set of data (T_s, T_{on}, and T_p) was obtained for constant heating rates ($\beta = dT/dt$) from which the kinetic parameters were calculated using the following equations:

$$\log \beta = a\frac{1}{T} + b \tag{3}$$

where β is the heating rate (K/min) and T is the temperature T_s, T_{on}, or T_p (K). By plotting log β against $1/T$, the effective activation energy (E_a) and the pre-exponential factor (A) were determined directly from the slope and intercept according to:

$$a = -0.4567\frac{E_a}{R} \tag{4}$$

$$b = -2.315 + \log\left(A\frac{E_a}{T}\right) \tag{5}$$

where a and b are the slope and intercept from equation (3), respectively, and R is the universal gas constant (8.31 J/mol K). Therefore, the effective activation energy (E_a) and the constant rate (k) are calculated from:

$$E_a = -2.19 R\frac{d\log \beta}{dT^{-1}} \tag{6}$$

$$k = A \exp^{\left(\frac{E_a}{RT}\right)} \tag{7}$$

5.3. Induction time

As mentioned earlier, the period of time where no change in the heat flow signal occurs is known as the induction time and its length is considered as a measurement of lipid stability [17]. The determination of the induction period is routinely conducted to evaluate stability of oils, lubricants, biodiesel, and pharmaceutical products [34-38]. Unfortunately, no

standard protocol for the determination of the oxidation induction time has been developed
for lipids.

Figure 7a illustrates the standard protocol for determination of oxidation induction time
(OIT). The polymer sample (15 mg) is rapidly heated under nitrogen atmosphere (≤
20°C/min) until it reaches the temperature that corresponds to time, t_1, which is the starting
point for OIT determination. At t_1, the atmosphere is switched to oxygen and the sample is
held at the same temperature until t_2 is reached. The difference between t_2 and t_1 is the OIT.
A disadvantage of this protocol is to find an adequate temperature for the isothermal stage.
For example, the use of a low temperature might considerably increase the OIT while the
use of a high temperature might oxidize the sample immediately, making difficult to obtain
a reliable baseline. In tests conducted in 16 different laboratories, it was demonstrated that
OIT is associated with a high degree of uncertainty. On the other hand, Figure 7b illustrates
the experimental protocol for the determination of oxidation induction temperature (OIT*).
In this case, the sample is continuously heated at a constant heat rate (for example,
12°C/min) under oxygen atmosphere.

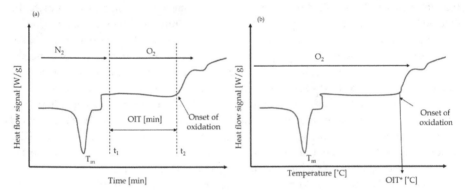

Figure 7. Determination of oxidation induction time, OIT (a), and oxidation induction temperature,
OIT* (b). T_m – melting temperature, t_1 – start of the oxidation induction time, t_2 – end of the oxidation
induction time. Adapted from reference [35].

6. Kinetic studies of oxidation of fats and oils

Tables 1 and 2 summarize oxidation studies of fats and oils using DSC. Most of the
isothermal studies were conducted between 80 and 180°C with different flow rates
varying from 10 to 100 mL/min. The amount of sample used also varied from 3 to 30 mg
and most of these studies used open pans. The oxidation onset times were quite diverse
(23-108 min) and therefore the E_a values ranged from 50 to 130 kJ/mol, depending mainly
on the oil composition, temperature, and amount of sample. For the non-isothermal
studies, the temperature ranged from 50 to 350°C and the heating rate used ranged from 1
to 25°C/min. The onset oxidation temperatures depended on the heating rate and oil
composition.

6.1. Isothermal studies

For isothermal oxidation, the heat flow signal generated at constant temperature is plotted against time (Fig 7a). From these curves, the start of the oxidation (t_1) and the end of the oxidation induction time (t_2) are first located and then used for analysis [39].

Table 1 summarizes the isothermal oxidation studies of fats and oils, such as peanut, safflower seed, blackcurrant seed, rice bran, cotton seed, Buriti seed, passion fruit seed, sunflower seed, soybean, linseed, canola seed, coconut, grape seed, palm seed, and sesame seed. Earlier studies on isothermal oxidation were imprecise because the baselines obtained were highly unstable, making it difficult to obtain oxidation onset times [40, 41]. However, stable baselines were currently obtained in the isothermal oxidation of linoleic, linolenic, oleic, stearic, peanut, safflower and blackcurrant seed oils [42]. These oils were heated under argon flow. After thermal equilibrium was reached, the gas flow was switched to oxygen, allowing the oxidation to start. The temperature used to conduct the oxidation test should be far below the self-ignition temperature for fats and oils (~ 350°C). Thus, the recorded thermal events are due to lipid oxidation rather than combustion. Using this approach, the obtained induction times were reproducible and highly influenced by the test temperature and sample composition. Attempting to validate the isothermal DSC oxidation, an earlier study [43] isothermally oxidized purpurea seed, rice bran and cotton seed oils using the DSC and the Rancimat methods. At each tested temperature, the DSC oxidation times were shorter than those obtained using the Rancimat method for the same oil. Despite these differences, the oxidation times were satisfactorily correlated with the Rancimat induction times. Similarly, longer induction times were obtained with the use of the Rancimat method for Buriti pulp seed oil, rubber seed oil and passion fruit seed oil [44]. In these studies, DSC rapidly reaches the threshold of the heat flow signal. This difference is attributed to the small sample used in DSC experiments, allowing a higher oil-oxygen ratio compared to the Rancimat method. Moreover, to oxidize the oil, DSC employs pure oxygen (99% purity) while the Rancimat method uses air (~21% of oxygen).

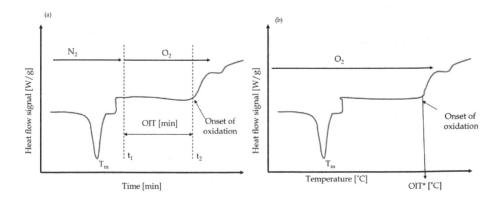

Lipid	Experimental protocol	Oxidation induction times (OIT, min)				Ref.
Linoleic, linolenic, peanut, oleic, stearic, safflower and blackcurrant seed	- 80-160°C - 5 mg in an open pan - Flow rate = 8.3 mL/min -Argon flow to equilibrate	At 120°C - Olive = 108 - Blackcurrant = 95 - Corn = 78 - Peanut = 68 - Sunflower = 69 - Linseed = 50 - Safflower = 43		At 130°C - Olive = 105 - Blackcurrant = 50 - Corn = 50 - Peanut = 54 - Sunflower = 22 - Linseed = 21 - Safflower = 18		[42]
Rice, cotton seed and *B. purpurea*	- 110, 120, 130 and 140°C - 5 mg in an open pan - Flow rate = 50 mL/min	At 110°C - *B. purpurea* = 483 - Rice = 132 - Cotton = 172	At 120°C - *B. purpurea* = 269 - Rice = 72 - Cotton = 92	At 130°C - *B. purpurea* = 99 - Rice = 36 - Cotton = 42	At 140°C - *B. purpurea* = 48 - Rice = 18 - Cotton = 20	[43]
Buriti, rubber seed and passion fruit oil	- 100, 110, 120, 130 and 140°C - 5 mg in an open pan - Flow rate = 50 mL/min	100°C - Buriti = 116 - Rubber = 106 - Passion = 778	110°C - Buriti = 42 - Rubber = 51 - Passion = 369	120°C - Buriti = 23 - Rubber = 25 - Passion = 163	130°C - Buriti = 7.8 - Rubber = 7.5 - Passion = 66.5 140°C - Buriti = 3.3 - Rubber = 3.3 - Passion = 30.3	[44]
Rapeseed and sunflower	- 120°C - 30 mg in a sealed pan - Flow rate = 600 mL/min	- Rapeseed: $OIT_{120°C}$ = 1.5-1.9 min - Sunflower: $OIT_{120°C}$ = 0.6-1.2 min				[45]
Linseed	- 130°C - 5 mg in a sealed pan - Flow rate = 60 mL/min	- Linseed: $OIT_{130°C}$ = 23.1 - Linseed + 0.05% of antioxidant blend = 25.1 - Linseed + 0.2% of antioxidant blend = 33.1 - Linseed + 0.01% Butylated hydroxyl anisole (BHA) = 24 - Linssed + 0.02% BHA = 30.1				[49]

Lipid	Experimental protocol	Kinetic parameters (E_a, kJ/mol; A min^{-1}; k, min)	Ref.
Rapeseed, soybean, corn and sunflower	- 121 to 149°C - 3.5-6.5 mg in an open pan - Flow rate = 100 mL/min	- Soybean: $OIT_{121-150°C}$ = 11-2.9; E_a = 67.2; A= 2.7x10^7 - Rapeseed: $OIT_{121-130°C}$ = 16.2-10.4; E_a = 71.9; A = 1.2x10^8 - Corn: $OIT_{114-130°C}$ = 23-3; E_a = 98.6; A= 5.1x10^{11} - Sunflower: $OIT_{106-130°C}$ = 11-2.8; E_a = 84.6; A= 1.4x10^{10}	[47, 48]
Soybean, rapeseed, and sunflower oil	- 110-150°C. - 3-4 mg in an open pan - Flow rate = 100 mL/min	- Soybean: $OIT_{110-150°C}$ = 139-11; E_a = 72.4; A= 6.3x10^7; $k_{150°C}$ = 0.073 - Rapeseed: $OIT_{110-150°C}$ = 122-14; E_a = 83.4; A= 1.6x10^9; $k_{150°C}$ = 0.084 - Sunflower: $OIT_{110-150°C}$ = 64-4.4; E_a = 86.2; A= 8.7x10^9; $k_{150°C}$ = 0.201	[50]
Canola, coconut, corn, grapeseed, peanut, palm kernel, palm olein, safflower, sesame and soybean	- 110-140°C. - 5 mg in an open pan - Flow rate = 50 mL/min	- Canola: $OIT_{110-140°C}$ = 259-37; E_a = 86; $k_{128°C}$ = 0.016 - Coconut: $OIT_{110-140°C}$ = 325-44; E_a = 86.9; $k_{128°C}$ = 0.011 - Corn: $OIT_{110-140°C}$ = 166-21.4; E_a = 88.1; $k_{128°C}$ = 0.021 - Grape seed: $OIT_{110-140°C}$ = 74-7.5; E_a = 99.9; $k_{128°C}$ = 0.056 - Peanut: $OIT_{110-140°C}$ = 127-12.3; E_a = 99.1; $k_{128°C}$ = 0.025 - Palm kernel: $OIT_{110-140°C}$ = 539-70; E_a = 89.4; $k_{128°C}$ = 0.007 - Palm olein: $OIT_{110-140°C}$ = 515-82; E_a = 79.9; $k_{128°C}$ = 0.006 - Safflower: $OIT_{110-140°C}$ = 88-8; E_a = 104.3; $k_{128°C}$ = 0.055 - Sesame: $OIT_{110-140°C}$ = 542-20; E_a = 88.8; $k_{128°C}$ = 0.007 - Soybean: $OIT_{110-140°C}$ = 124-20; E_a = 80.8; $k_{128°C}$ = 0.029	[51, 52]
Soybean, rapeseed, corn and peanut oil	- 108-162°C. - 2-5 mg in an open pan. - Flow rate = 100 mL/min.	- Rapeseed: $OIT_{130-150°C}$ = 33-11; E_a = 64.4 - Soybean: $OIT_{138-162°C}$ = 8.9-2.3; E_a = 51.3 - Corn: $OIT_{147-160°C}$ = 13.5-5.7; E_a = 77.5 - Palm: $OIT_{108-121°C}$ = 24-6.5; E_a = 82.3	[53]
Lauric, myristic, palmitic, stearic acids and their ester	- 155-180°C. - 5-8 mg in an open pan. - Flow rate = 100 mL/min.	- Palmitic: E_a = 125.1 ± 11.2 - Ethyl palmitate: E_a = 126.6 ± 5.0 - Glycerol tripalmitate: E_a = 105.3 ± 7.7 - Stearic acid: E_a = 134.3 ± 12.0 - Ethyl stearate: E_a = 128.5 ± 2.6 - Glycerol tristearate: E_a = 102.5 ± 10.9 - Lauric acid: E_a = 97.3 ± 7.3 - Ethyl laurate: E_a = 127.3 ± 6.3 - Ethyl myristate: E_a = 117.1 ± 5.4	[54]
Blends of cocoa butter/cocoa butter fat like	- 130, 140, 150, and 160°C. - 5 mg in an open pan. - Flow rate = 100 mL/min.	Cocoa butter/cocoa butter fat like - 100/0: E_a = 106.2; A = 5.6x10^{10}; $k_{160°C}$ = 0.008 - 90/10: E_a = 105.1; A = 4.4x10^{10}; $k_{160°C}$ = 0.009 - 80/20: E_a = 105.9; A = 6.7x10^{10}; $k_{160°C}$ = 0.011 - 70/30: E_a = 95.1; A = 3.9x10^9; $k_{160°C}$ = 0.013 - 60/40: E_a = 99.5; A = 1.5x10^{10}; $k_{160°C}$ = 0.015 - 50/50: E_a = 104.4; A = 6.7x10^{10}; $k_{160°C}$ = 0.017 - 0/100: E_a = 120.5; A = 1.9x10^{13}; $k_{160°C}$ = 0.055	[55]

Table 1. Summary of isothermal oxidation studies of lipids using differential scanning calorimetry

A comparative study of sunflower seed oil and rapeseed oil oxidation using DSC and volatile analysis showed that the ratio of hexanal/2-trans-nonenal linearly correlates with the onset heat flow signal of the DSC spectra [44]. In addition, a correlation between the peroxide values with the oxidation onset time was developed to monitor rapeseed oil oxidation [44]. Unfortunately, the correlation was valid only for oils with peroxide values lower than 30 mmol O_2/kg oil. In other study, the experimental oxidation data obtained from electron spin spectroscopy was compared with the data obtained from DSC oxidation [46]. The onset oxidation times were highly correlated for different fat and oil blends. However, the obtained correlations were valid only at moderate temperatures of 60°C, which considerably limits the applicability of electron spin spectroscopy for oxidation analysis.

An investigation of isothermal oxidation of oils (e.g. rapeseed, soybean, corn and sunflower oils) used onset time (t_{on}) to rank the oxidized oils in terms of their oxidative stability [47]. The high maximum heat flow time (t_p) value indicates that the oil is more stable. Although these relationships were statistically validated, a single parameter to evaluate the oxidative stability can lead to overestimation of the oxidative stability. In the same study, the authors considered t_p to be proportional to the rate of oxidation, which might not be valid. Indeed, peroxide value determinations showed that t_p represents the oxidation termination stage while t_{on} is associated with the rate of initiation [48]. Furthermore, the addition of antioxidants prolongs only t_{on} while t_p values were minimally affected [48]. The same behavior was also observed in the isothermal oxidation of linseed oil with the use of antioxidants [49]. The addition of BHA (butylated hydroxyl anisole) and a mixture of antioxidants (tocopherol, ascorbyl palmitate, citric acid and ascorbic acid) prolonged the onset time at 130°C. All these approaches provided information of great value for validation of the DSC isothermal oxidation. However, these equations are limited for a set of temperatures and specific oils (Table 1). Additional factors, such as degree of saturation, amount of free fatty acids, chain length and the presence of natural antioxidants were not considered.

Using the isothermal method, the sample is heated at a constant temperature and the released heat is recorded as a function of temperature. Such situation allows the identification of the maximum heat flow time (t_m), which linearly correlates with the temperature [39,48, 49].

$$\log t_m = A \cdot T^{-1} + B \tag{8}$$

where A and B are regression parameters. Due to the excess of oxygen generated by a constant flow rate, the formation of peroxides is considered to be independent of the oxygen concentration, which also means that the autoxidation is a first order reaction [48, 49]. This is an essential assumption for the calculation of kinetic parameters, such as effective activation energy (E_a), pre-exponential factor (A), and reaction rate (k).

$$E_a = 2.19 \cdot R \cdot \frac{d \log(t_m)}{dT^{-1}} \tag{9}$$

Equations 8 and 9 have been applied not only to obtain the maximum heat flow signal [47-49] but also to obtain the oxidation onset time. An attempt to correlate induction times and kinetic parameters of isothermally heated rapeseed, soybean and sunflower seed oils at different temperatures was proposed [50]. The authors correlated the oxidation induction times with the tested temperature using Arrhenius-like equations. The obtained kinetic parameters were comparable to those obtained by the Rancimat method. The onset time values of 12 different oils obtained by DSC were reduced by half of their previous values for every increase of 10°C in the oxidation temperature [51]. These relationships were further used to obtain kinetic parameters of DSC oxidation [52]. The E_a values of rapeseed, soybean, corn and peanut oils were strongly influenced by the amount of saturated fatty acids [53]. The oxidation of saturated fatty acids (C_{12}-C_{18}) and their esters revealed that the E_a values (100-125 kJ/mol) were within the same range for all the tested oils [54]. This suggests that the isothermal oxidation is not influenced by the carbon chain length. Another important conclusion from this investigation [54] is that the start of oxidation is similar for fatty acids, their esters and triglycerides. The kinetic oxidation parameters of cocoa butter blends were obtained using the oxidation onset time [55]. The blends were oxidized from 130 to 160°C in an open pan. Interestingly, the E_a and A values were slightly affected by the addition of saturated fatty acids. But, k values considerably changed with the amount of saturated fatty acids. Thus, k values can be used to rank the oxidative stability of cocoa butter blends. However, the use of k values to evaluate oxidative stability might be valid only at the temperature tested since changes in the reaction mechanisms can occur as a function of temperature.

6.2. Non-isothermal studies

For non-isothermal oxidation studies, two maximum heat flow peaks are commonly observed in the DSC spectra [56-59]. But, only the first peak is related to lipid oxidation. This was demonstrated in non-isothermal oxidation studies of corn and linseed oils with different peroxide values [57]. A decrease in the onset temperature was observed as the peroxide value increased. Contrary, the first and second peak temperatures were not affected by increasing the peroxide value. Consequently, the first peak can be related to hydroperoxides formation while the second peak can be due to further oxidation of peroxides. In an earlier study [58], the weight loss of lecithin during non-isothermal heating under nitrogen flow rate was analyzed. The thermogravimetric analysis showed that in the temperature range of the first peak only 4% of weight was lost but above that temperature, the weight loss considerably increased. Therefore, changes in the DSC signal within the range of the first peak temperature were attributed to oxidation and those changes above that temperature corresponded to thermal degradation rather than oxidation (Figure 8).

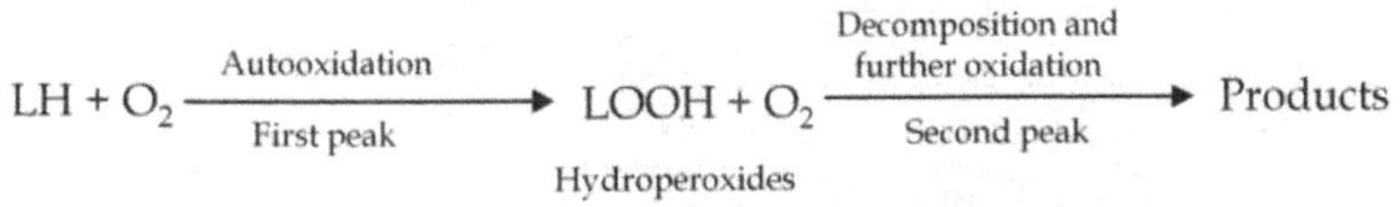

$$LH + O_2 \xrightarrow[\text{First peak}]{\text{Autooxidation}} \underset{\text{Hydroperoxides}}{LOOH + O_2} \xrightarrow[\text{Second peak}]{\substack{\text{Decomposition and} \\ \text{further oxidation}}} Products$$

Figure 8. Reaction mechanism proposed for lipid oxidation under non-isothermal conditions.

The proposed reactions in Figure 8 resemble to an autocatalytic reaction scheme. Indeed, computer simulated DSC oxidation curves using autocatalytic scheme fitted well the experimental data [56]. Similarly, the non-isothermal oxidation of mustard oil was best described by an autocatalytic reaction scheme [58]. Therefore, it was proposed that T_{on} is the most representative reference point of lipid oxidation under non-isothermal conditions. In addition, kinetic parameters calculated from T_{on} were comparable to those obtained at isothermal DSC conditions (Tables 1 and 2) [56-59]. Similarly, non-isothermal kinetics was used to evaluate the oxidative stability of commercial olive oil samples [59-62]. The obtained kinetic parameters were comparable to those obtained with the Rancimat method.

Lipid	Experimental protocol	Kinetics parameters (E_a, kJ/mol; A min^{-1}; k, min)	Ref.
Corn and linseed oil	- 50-300°C - 5 mg in an open pan - β = 5-20°C/min - Flow rate = 166 mL/min	- Corn: E_a = 69.5±9.7; A = 1.25x10^8 - Linseed: E_a = 132.7±8.8; A = 1.15x10^{16} *Onset temperature* - Corn + 47 mmol O_2/kg = 146°C - Corn + 70 mmol O_2/kg = 142°C - Corn + 93 mmol O_2/kg = 137°C - Corn + 136 mmol O_2/kg = 137°C - Corn + 145 mmol O_2/kg = 128°C - Corn + 158 mmol O_2/kg = 127°C - Linseed + 31 mmol O_2/kg = 144°C - Linseed + 119 mmol O_2/kg = 131°C - Linseed + 180 mmol O_2/kg = 129°C - Linseed + 252 mmol O_2/kg = 125°C - Linseed + 349 mmol O_2/kg = 117°C - Linseed + 383 mmol O_2/kg = 121°C	[56]
Linolenic acid and soy lecithin	- 50-300°C - 5 mg in an open pan - β = 2-20°C/min - Flow rate = 6600 mL/min	- Linolenic: E_a from T_{on}: 65±4; A = 2.3x10^7; $k_{100°C}$ = 0.016 - Linolenic: E_a from T_p: 78.9±6.9; A = 9.9x10^7 - Lecithin: E_a from T_{on}: 97±8l; A = 2.4x10^{11}; $k_{100°C}$ = 0.001 - Linolenic: E_a from T_p: 141.4±4; A = 9.4x10^{15}	[57]
Mustartd oil	- 140-350°C - 3-5 mg in an open pan - β = 2, 5, 7.5, 10, and 15°C/min - Flow rate = 100 mL/min	- From onset: E_a = 90.6; A = 3.4x10^9; $k_{225°C}$ = 0.97 - From 1st peak: E_a = 88.5; A = 1.4x10^9; $k_{225°C}$ = 0.90 - From 2nd peak: E_a = 84.6; A = 4.4x10^7; $k_{225°C}$ = 1.12	[58]
Olive oil	- 2-3 mg in an open pan - β = 4, 5, 7.5, 10, 12.5, 15°C/min - Flow rate = 100 mL/min	- E_a calculated from T_{on}: 72-104; A = 3.31x10^8 -1.1x10^{12} ; $k_{120°C}$ = 0.09-0.015	[59]
Blends of soybean/AMF	- 100 to 350°C - 5-15 mg in an open pan - β = 2.5-12.5°C/min - Flow rate = 100 mL/min	*Soybean/AMF* - 100/0: E_a = 93.5; A = 1.25.6x10^{10}; $k_{200°C}$ = 0.57 - 90/10: E_a = 59.5; A = 2.5x10^6; $k_{200°C}$ = 0.68 - 80/20: E_a = 58.4; A = 2.1x10^6; $k_{200°C}$ = 0.71 - 70/30: E_a = 64.5; A = 1.1x10^7; $k_{200°C}$ = 0.80 - 60/40: E_a = 70.2; A = 5.2x10^7; $k_{200°C}$ = 0.88 - 50/50: E_a = 73.2; A = 1.1x10^8; $k_{200°C}$ = 0.95 - 40/60: E_a = 102.7; A = 2.6x10^{11}; $k_{200°C}$ = 1.18 - 30/70: E_a = 102.8; A = 2.7x10^{11}; $k_{200°C}$ = 1.20 - 20/80: E_a = 105.8; A = 5.3x10^{11}; $k_{200°C}$ = 1.26 - 10/90: E_a = 117.4; A = 1.4x10^{13}; $k_{200°C}$ = 1.55 - 0/100: E_a = 89.5; A = 8.4x10^9; $k_{200°C}$ = 1.09	[62]
Oleic, erucic, linoleic, linolenic and their ethyl esters and glycerol trioleate and trilinoleate	- Temperature = 50-300°C - 5 mg in an open pan - β = 2-20°C/min - Flow rate = 600 mL/min	- Erucic: E_a = 89.6±4.4; A = 4.9 x10^{10}; $k_{90°C}$ = 0.006 - Oleic: E_a = 88.4±4.7; A = 1.0 x10^{11}; $k_{90°C}$ = 0.021 - Oleate: E_a = 95±4.7; A = 9.4 x10^{10}; $k_{90°C}$ = 0.002 - Trioleate: E_a = 91.8±13.3; A = 4.4 x10^{10}; $k_{90°C}$ = 0.002 - Linoleic: E_a = 72±2.9; A = 1.6 x10^9; $k_{90°C}$ = 0.071 - Linoleate: E_a = 76.4±5; A = 2.4x10^9; $k_{90°C}$ = 0.024 - Trilinoleate: E_a = 74.3±3; A = 9.6x10^8; $k_{90°C}$ = 0.020 - Linolenic: E_a = 62.4±3.7; A = 2.6x10^8; $k_{90°C}$ = 0.027 - Linolenate: E_a = 74.5±8.2; A = 4.2x10^9; $k_{90°C}$ = 0.082	[63, 64]
AMF with low, medium and high CLA content	- 100 to 350°C - 2 mg in an sealed pan - β = 3, 6, 9, 12 and 15°C/min - Flow rate = 50 mL/min	- Low CLA: E_a = 146.5; A = 4.1x10^{14}; $k_{200°C}$ = 0.026 - Med CLA: E_a = 112.4; A = 3.6x10^{10}; $k_{200°C}$ = 0.014 - High CLA: E_a = 87.6; A = 6.3x10^7; $k_{200°C}$ = 0.013	[9]
Cotton, corn, canola, safflower, high oleic safflower, high	- 1-1.5 mg in a sealed pan - β = 1, 5, 10, 15 and 20°C/min	- Cotton: E_a = 63.3; A = 9.2x10^6; k = 0.37 - Corn: E_a = 77.7; A = 2.4x10^8; k = 0.43	[7]

Lipid	Experimental protocol	Kinetics parameters (E_a, kJ/mol; A min^{-1}; k, min)	Ref.
linoleic safflower, high oleic sunflower, soybean and sunflower		- Canola: E_a = 88.4; A = 7.6x10^9; k = 0.51 - Safflower: E_a = 75.2; A = 1.8x10^8; k = 0.44 - High oleic safflower: E_a = 88.7; A = 3.1x10^9; k = 0.48 - High linoleic safflower: E_a = 73.5; A = 1.1x10^8 k = 0.42 - High oleic sunflower: E_a = 86.5; A = 1.8x10^9; k = 0.47 - Soybean: E_a = 79.6; A = 4.1x10^8; k = 0.44 - Sunflower: E_a = 63.8; A = 1.1x10^7; k= 0.38	
Base oil lubricants	- 0.5 mg in a sealed pan - β = 5, 10 and 20°C/min - Flow rate = 100 mL/min	- E_a = 13.8-83.9 kJ/mol; k = 0.11-0.75 min^{-1}.	[25]
Rapeseed, soybean, sunflower, lard and highly rancid oils	- 100-360°C - 3-4 mg in an open pan - β = 5-20°C/min - 166 mL/min - Antioxidants: BHT, BHA and PG	- Rapeseed: E_a = 64-73; A = 7.4x10^5-4.4x10^6 - Soybean: E_a = 62-64; A = 6.1x10^5-7.1x10^5 - Sunflower: E_a = 60-62; A = 4.1x10^5-6.6x10^5 - Lard: E_a = 92-93; A = 3.5x10^8-2.7x10^8	[65]
Fat extracted from baby formulas	- 3-4 mg - β = 4, 6, 7.5, 10, 12.5, 15°C/min - Flow rate = 50 mL/min	- E_a = 80-106, A = 7.4x10^9-1.1x10^{-13}; $k_{120°C}$ = 0.079-0.122	[19]
High oleic sunflower (HOS) and castor oil with blends of antioxidants	- 100 to 250°C - 3-3.3 mg in an open pan - β = 10°C/min - Flow rate = 10 mL/min - Antioxidants (0.5-2 wt %): TOC, PG, AA and MBP	- HOS: OOT = 191°C - HOS + PG: OOT = 225-233°C - HOS + AA: OOT = 206°C - HOS + TOC: OOT = 190-214°C - HOS + MBP: OOT = 226-242°C - Castor: OOT = 197 °C - Castor + PG: OOT = 233-242°C - Castor + AA: OOT = 208-212°C - Castor + TOC: OOT = 195-203°C - Castor + MBP: OOT = 212-234°C	[66]
Linolenic acid (LNA) with different phenols	- 50-300°C - 3-5 mg in an open pan - β = 5, 10, 20, 40, and 80°C/min - Flow rate = 250 mL/min	- LNA: E_a = 70.4; A = 5.6x10^9 min^{-1}; $k_{90°C}$ = 0.792 - 0.26 BHT/LNA: E_a = 73±3.7; A = 8.9x10^9; $k_{90°C}$ = 0.541 - 0.82 BHT/LNA: E_a = 79.6±8.4; A = 2.4 x 10^{10}; $k_{90°C}$ = 0.17 - 1.51 BHT/LNA: E_a = 87±5.5; A = 2.0x10^{11}; $k_{90°C}$ = 0.134 - 3.0 BHT/LNA: E_a = 91.9±4.5; A = 6.5x10^{11}; $k_{90°C}$ = 0.091 - 4.0 BHT/LNA: E_a = 95.8±2.4; A = 2.1x10^{12}; $k_{90°C}$ = 0.049	[67]
LNA with BHT, olivetol and DHZ	- 50-350°C - 5 mg in an open pan - β = 2-20°C/min - Flow rate = 250 mL/min	- LNA: E_a = 69.8±7.8; A = 2.2x10^9 - 1.2-11.6 BHT/LNA: E_a = 82-141 - 1.1-8.2 DHZ/LNA: E_a = 88-187 - 1.1-5.6 Olivetal/LNA: E_a = 84-129	[68]

AA – ascorbic acid; TOC – tocopherol; MBP – methylenebis (2,6 ditert-butylphenol); PG – propyl gallate; BHT- Butylated hydroxytoluene, DHZ-dehydrozingerone, AMF-anhydrous milk fat, CLA-conjugated linoleic acid, OOT – oxidation onset temperature.

Table 2. Summary of non-isothermal oxidation studies of lipids using differential scanning calorimetry

According to the Arrhenius principle, oil with a high E_a value oxidizes faster at high temperatures, while oil with a low E_a value oxidizes faster at low temperatures. Unfortunately, calculated values of E_a should not be used as a single parameter to rank the oxidative stability of lipid systems. This was exemplified in blends of soybean/anhydrous milk fat [63] that were non-isothermally oxidized. Interestingly, as the percentage of unsaturated fatty acids increased, the onset temperature of oxidation decreased and the only kinetic parameter that exhibited the same pattern was the constant rate of oxidation. The calculated E_a value is the cumulative effect of all the E_a values available in the system during oxidation, including intermediate compounds that have their own kinetic values. An equation representing the overall activation energy for autoxidation of lipids was earlier proposed [54, 56]. The overall effect included activation energies of initiation (E_i), propagation (E_p) and termination (E_t) based on the classical rate equation for autoxidation of hydrocarbons.

Unfortunately, equation (10) has been applied to limited fatty acids (C_{12}-C_{18}) [56, 57] and correlations between other kinetic parameters and the initiation and termination activation energies are needed.

$$E = E_p + \frac{1}{2}E_i - \frac{1}{2}E_t \tag{10}$$

The non-isothermal oxidation of different fatty acids and their esters [56, 57] showed that the calculated E_a values are similar among the different tested samples, indicating that the oxidation does not occur on free or esterified carboxyl groups of fatty acids. The non-isothermal oxidation of anhydrous milk fat with different ratios of unsaturated/saturated fatty acids showed that the start temperature of oxidation shifted to lower values as the ratio increased [9]. More importantly, the kinetics parameters (E_a, A and k) calculated also decreased. The onset temperature of oxidation not only is affected by the amount of saturated fatty acids but also by the presence and abundance of aromatic compounds and their alkyl substitutions [7, 25]. Kinetic parameters obtained from different oils were compared with structural parameters obtained with NMR spectroscopy. Moreover, an increase in the methylene carbons of the fatty acid chains increased the oxidative stability while conjugated structures were rapidly oxidized.

The addition of antioxidants to enhance oxidative stability of oils can be evaluated using the DSC in non-isothermal mode. A novel approach based on DSC data, named protective factor (PF) was provided in the literature [65]. In this investigation, the oxidation of methyl esters derived from rapeseed and waste frying oil was monitored at different concentrations of BHT and pyrogallol (PG). The oxidation onset temperature asymptotically increased with the addition of BHT and PG, without finding an optimum antioxidant concentration. In addition, the increase in the heating rate might change the reaction mechanisms in which antioxidants can capture free radicals, making difficult to compare their effectiveness. Therefore, the protective factor concept [65] was developed according to:

$$Protective\ factor\ (PF) = \frac{onset\ temperature\ of\ oil\ with\ antioxidant}{onset\ temperature\ of\ oil\ without\ antioxidant} \tag{11}$$

Values of PF lower than 1 means that the antioxidant has a pro-oxidant effect. On the other hand, PF values greater than one can be considered as a measurement of antioxidant effectiveness. Another important factor is the physical stability of the antioxidants. In a comparative study [65, 68], the addition of BHT, BHA and PG was evaluated in rapeseed, soybean, sunflower and lard oils. BHA and BHT were not effective antioxidants due to their volatility. These antioxidants escaped from the heated oil before they can react to neutralize free radicals. An optimum antioxidant concentration was found for BHT (8.4 mmol), BHA (2.8 mmol) and olivetol (4.5 mmol) in oxidized linolenic acid [63-65]. After the optimal concentration, a decrease in the antioxidant activity was observed. Interestingly, at high temperatures (<180°C), the antioxidants are no longer stable and their effectiveness significantly decreased. Similarly, α-tocopherol and L-ascorbic acid 6-palmitate were not effective even at high concentrations (up to 2% wt) during the oxidation of either high oleic sunflower oil or castor oil [66, 69].

6.3. Compensation theory

A proper kinetic interpretation should include the compensation effect, which is usually used to explain whether the variations on effective activation energy values have physical meaning or they are caused by either variations of process conditions or complexity of the reaction systems. In previous investigations [69, 70], the compensation effect is illustrated for various thermal degraded materials, such as polymers, cellulosic materials and $CaCO_3$. The compensation effect can be evaluated by plotting $\ln k$, obtained by equation (7), against $1/T$. Figure 9 shows that an increase in the effective activation energy causes an increase in $\ln k$ (equation 7). Similarly, a decrease in the effective activation energy results in a lower value of $\ln k$. The point of concurrence, where the different lines intercept, corresponds to $\ln k_{iso}$ and $1/T_{iso}$ (k_{iso} is the isokinetic rate constant and T_{iso} is the isokinetic temperature), which indicates the existence of the compensation effect.

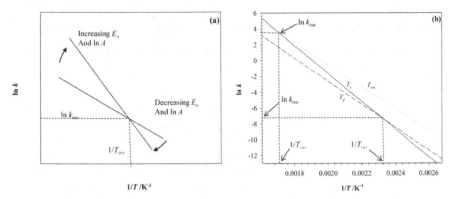

Figure 9. Arrhenius plot ($\ln k$ vs $1/T$) for the non-isothermal oxidation: (a) compensation effect theory adapted from reference [69], and (b) anhydrous milk fat [9].

In studies of non-isothermal oxidation of anhydrous milk fat (AMF) [9], there is no concurrence at a single point. Therefore, the non-isothermal oxidation of AMF does not exhibit a compensation effect and thus the variations in kinetic parameters have no physical background. In fact, AMF is a very complex fat mainly composed of triacylglycerols that have a glycerol backbone to which three fatty acid moieties are esterified. These triacylglycerols are extremely diverse in chain lengths, position and number of unsaturations of their fatty acids [72]. Moreover, more than 400 fatty acids in milk fat were found [73]. Therefore, several reactions with different constant rates simultaneously occur and DSC only detects those reactions that have the greatest exothermal effect. This could explain why the variations in effective activation energy values have no physical meaning or there is no compensation effect.

The isokinetic temperature can also be defined as the temperature at which two rate constants of two different reactions are equal. For a set of kinetic parameters of two different reactions, the isokinetic temperature can be expressed as:

$$A_1 \cdot exp^{\left(\frac{-E_1}{R \cdot T_{iso}}\right)} = A_2 \cdot exp^{\left(\frac{-E_2}{R \cdot T_{iso}}\right)} \tag{12}$$

The isokinetic temperature of autoxidation of lecithin and linolenic acid calculated using equation (12) was 167°C [57]. This observation exemplifies the difficulty in determining the oxidative stability of multicomponent systems. For example, if the oxidation test is conducted at a temperature below T_{iso}, linolenic acid oxidizes faster than lecithin. But, if the same test is conducted at a temperature equal to T_{iso}, both lipids have the same oxidative stability. Contrary, lecithin oxidizes faster than linolenic acid above T_{iso}. Consequently, the estimation of the oxidative stability based only on the onset temperature is misleading. Therefore, interpretation of the shape of non-isothermal oxidation curves in combination with kinetic parameters can provide a better interpretation of the oxidative behavior in multicomponent systems.

7. Case studies of lipid oxidation after processing using new technologies

7.1. Kinetics of non-isothermal oxidation of anhydrous milk fat (AMF) rich in conjugated linoleic acid (CLA) after hydrolysis

AMF is the richest source of CLA composed of geometrical and positional isomers of linoleic acid. CLA has potential benefits, such as cancer prevention, atherosclerosis, weight control, and bone formation [74]. Additionally, CLA concentration in milk can be markedly enhanced through diet manipulation and nutritional management of dairy cattle [75]. In CLA-enriched AMF, CLA is distributed throughout different triacylglycerols together with other fatty acids, limiting its applicability as ingredient in different milk fat-based products. One known approach to produce free fatty acids (FFA) is through enzymatic hydrolysis. The enzymatic hydrolysis of AMF rich in CLA yielded around 88% of free fatty acids (FFA) [76, 77]. Unfortunately, FFAs are more susceptible to oxidation than those fatty acids attached to the triacylglycerol backbone.

Figure 10 shows the DSC curves of non-hydrolyzed and hydrolyzed AMF rich in CLA. Although the DSC curves were quite different between non-hydrolyzed AMF (blue and black lines) and hydrolyzed AMF (green and red lines), the maximum peak temperatures were in the same range. This observation supports the hypothesis that changes in the DSC signal below the temperature of the first peak can be attributed to oxidation and changes in DSC signal above T_P correspond to thermal decomposition and advanced oxidation products as shown in Figure 8. Figure 10 also shows that the oxidation of hydrolyzed AMF starts at low temperatures (~108°C), making difficult to obtain a consistent baseline. Therefore, the kinetic parameters were calculated using the onset temperature obtained as described in Section 5.1.

Table 3 shows the onset oxidation temperature and kinetic parameters of non-hydrolyzed and enzymatic hydrolyzed AMF rich in CLA. As expected, the E_a for the hydrolyzed fat is greater than that obtained for the non-hydrolyzed fat. Interestingly, the constant reaction

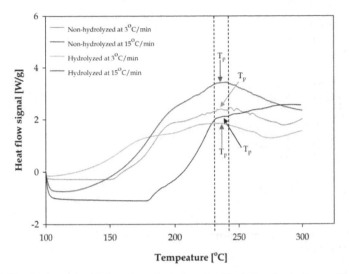

Figure 10. Non-isothermal oxidation of non-hydrolyzed and hydrolyzed anhydrous milk fat rich in
CLA. T_P - maximum heat flow temperature.

Heating rate [˚C/min]	Non-hydrolyzed AMF [°C]	Hydrolyzed AMF [°C]
3	173.47 ± 0.1	125.06 ± 1.4
6	184.71 ± 2.3	132.46 ± 0.3
9	192.13 ± 0.7	133.25 ± 1.2
12	197.85 ± 3.7	132.46 ± 0.3
15	198.66 ± 0.3	136.78 ± 0.1
E_a [kJ/mol]	82.42	175.78
A [min^{-1}]	8.7×10^7	2.3×10^{20}
$k_{130˚C}$ [min]	0.002	0.003
$k_{200˚C}$ [min]	0.075	10.28

Table 3. Onset temperature of oxidation for anhydrous milk fat (AMF)

rates (k) calculated at 130 °C were similar for the non-hydrolyzed and hydrolyzed samples.
However, the constant rate of hydrolyzed fat was dramatically higher than that of non-
hydrolyzed fat at 200°C.

The isokinetic temperatures of these two fats were calculated mathematically and
graphically using equation (12) and Arrhenius plot, respectively. Figure 11 shows the
compensation effect for the oxidation of non-hydrolyzed and hydrolyzed AMF rich in CLA.
From this figure, the T_{iso} and k_{iso} were obtained (T_{iso} = 120°C and k_{iso} = 0.0011 min). Similarly,
the T_{iso} and k_{iso} calculated mathematically with equation (12) were 118°C and 0.0010 min,
respectively. However, T_{iso} for non-hydrolyzed AMF is unrealistic since the start
temperature of oxidation is around 155°C. Therefore, oxidation of either non-hydrolyzed or
hydrolyzed AMF rich in CLA does not exhibit a compensation effect.

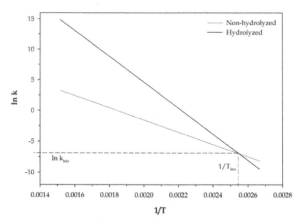

Figure 11. Arrhenius plot for the non-isothermal oxidation of non-hydrolyzed and hydrolyzed anhydrous milk fat rich in CLA.

7.2. Oxidative stability of AMF rich in CLA treated with pressure assisted thermal processing

CLA is not stable upon thermal processing and significant losses of its biological activity occurred through oxidation [78]. The application of high pressure (100-600 MPa) to a preheated sample can preserve the biological activity of functional compounds [79]. This is because the rise in temperature due to adiabatic heating is used to reach the target temperature, reducing the thermal damage due to the lack of temperature uniformity that occurs in traditional thermal processes [80, 81]. This technology is known as pressure-assisted thermal processing (PATP). Pressure alters interatomic distance, acting mainly on those weak interactions which bond energy is distance–dependent, such as van der Waals forces, electrostatic forces, hydrogen bonding and hydrophobic interactions of proteins. Based on the distance dependence, any pressurized sample would have its covalent bonds intact. This has been the central hypothesis in preserving the biological activity of functional compounds, such as ascorbic acid, folates, vitamins and anthocyanins [82-85].

The effects of PATP conditions on the antiradical ability of CLA in AMF were reported [86]. CLA can donate hydrogen to form a CLA-free radical that further reacts to inhibit hydroperoxides formation, depending on the final CLA retention. This suggests that the retained CLA after PATP treatment might enhance the oxidative stability of AMF. After PATP treatments, samples of AMF rich in CLA were oxidized at 6°C/min to calculate the start temperature of oxidation using the method described earlier in Section 5.1 of this chapter.

Figure 12 shows the influence of the retained CLA on the start temperature of oxidation (T_s). The stability of AMF is influenced in a non-linear mode by the CLA retention. Values of the start temperature of oxidation (~165-155°C) were obtained when the CLA retention was between 85 to 100%. On the other hand, the lowest value of the start temperature of oxidation (141°C) was obtained at 40% of CLA retention.

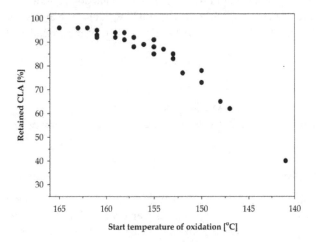

Figure 12. Influence of CLA retention on the start temperature (T_s) of oxidation in anhydrous milk fat rich in CLA.

A possible reason for the CLA-T_s relationship is that CLA can act as an antioxidant capturing those free radicals responsible for the lipid oxidation. This antioxidant behavior was clearly demonstrated in an earlier study by inducing lipid oxidation of fish oil by adding tert-butyl hydroperoxide. CLA effectively reduces lipid peroxidation as measured by chemiluminescence [86]. In addition, the antiradical or scavenging ability of CLA isomers measured by DPPH (2,2-diphenyl-1-picryl-hydrazyl-hydrate) technique depended on the CLA concentration [87].

8. Conclusions

The use of DSC to analyse lipid oxidation is a reliable, simple and convenient technique. It provides qualitative and quantitative information and offers unique advantages, such as the small amount of sample use, short test time, and good reproducibility. Additionally, the effectiveness of a particular antioxidant can be evaluated using DSC, measuring changes in the oxidation onset times or temperatures. The data obtained from DSC lipid oxidation have been extensively studied and correlated with other oxidation methods, such as the Rancimat method, PV, spectrophotometric and GC analysis of fats and oils from various sources.

A proper interpretation of DSC oxidation experiments should include sample composition, kinetic parameters (E_a, A, and k) in combination with the compensation theory. In multicomponent systems, the fat sometimes needs to be extracted from the matrix. Therefore, the lipid oxidative behaviour might be different from its original matrix. The results obtained from DSC oxidation depend upon the conditions used to prepare the sample and the heating protocol used. Factors, such as degree of saturation, amount of free fatty acids, chain length and the presence of natural antioxidants influence the oxidative stability and kinetic parameters. DSC can be coupled with other analytical techniques, such as GC, NMR, HPLC, etc, to provide a better description of the oxidative stability of fats and oils.

Author details

M.D.A. Saldaña and S.I. Martínez-Monteagudo
Department of Agricultural, Food and Nutritional Science, University of Alberta, Edmonton, AB, Canada

Acknowledgement

The authors thank to Alberta Livestock and Meat Agency Ltd. (ALMA) and to the Natural Sciences and Engineering Research Council of Canada (NSERC) for funding this project. Martinez-Monteagudo expresses his gratitude to Consejo Nacional de Ciencia y Tecnologia (CONACYT, Mexico) and Instituto de Inovacion y Transferencia Tecnologica (I^2T^2, Mexico) for the financial support (nr 187497).

9. References

[1] Kamal-Eldin A, Pokorn J (2005) Lipid oxidation products and methods used for their analysis. In Kamal-Eldin A, Pokorny J, editors. Analysis of lipid oxidation. AOCS Publishing, eBook ISBN: 978-1-4398-2239-5.

[2] O'Connor, TP, O'Brien NM (2006) Lipid oxidation. In Fox, PF, McSweeney P, editors. Advanced Dairy Chemistry, Volume 2: Lipids (3rd Edition). Springer-Verlag. pp. 557-600.

[3] Gray JI (1978) Measurement of lipid oxidation: A review. J. Am. Oil Chem. Soc. 55:539-546.

[4] Bradley DG, Min DB (1992) Singlet oxygen of foods. Crit. Rev. Food Sci. Nutr. 31:211-236.

[5] Shahidi F, Zhong Y (2010) Lipid oxidation and improving the oxidative stability. Chem. Soc. Rev. 39:4067-4079.

[6] Labuza TP (1971) Kinetics of lipid oxidation. Crit. Rev. Food Technol. 2:355-404.

[7] Adhvaryu A, Erhan SZ, Perez JM (2000) Oxidation kinetic studies of oils derived from unmodified and genetically modified vegetables using pressurized differential scanning calorimetry and nuclear magnetic resonance spectroscopy. Thermochim. Acta. 364: 97-97.

[8] Privett,OS, Blank ML (1962) Initial stages of autoxidation. J. Am. Oil Chem. Soc. 39: 465-468.

[9] Martínez-Monteagudo SI, Saldaña MDA, Kennelly JJ (2012) Kinetics of non-isothermal oxidation of anhydrous milk fat rich conjugated linoleic acid using differential scanning calorimetry. J. Therm. Anal. Calorim. 107: 973-981.

[10] Choe E, Min DB (2006) Chemistry and reactions of reactive oxygen species in foods. Crit. Rev. Food Sci. 46: 1-22.

[11] Brimberg UI, Kamal-Eldin A (2003) On the kinetics of the autoxidation of fats substrates with conjugated double bonds. Eur. J. Lipid Sci. Tech. 105: 17-22.

[12] Hamalainen TI, Sundberg S, Makkinen M, Hase T, Hopia A (2001) Hydroperoxide formation during autoxidation of conjugated linoleic acid methyl ester. Eur. J. Lipid Sci. Tech. 103: 588-593.

[13] Kenar JA, Gunstone FD, Knothe G (2007) Chapter 8: Chemical properties. In Harwood JL, Frank D, Gunstone FD, Dijkstra AJ, editors. The lipid handbook (3rd Edition). CRC Press. pp. 535-590.

[14] Lopez C, Lavigne F, Lesieur P, Bourgaux C, Ollivon M (2001) Thermal and structural behavior of milk fat. 1. Unstable species of anhydrous milk fat. J. Dairy Sci. 84: 756-766.

[15] Jensen RG (2002) The composition of bovine milk lipids: January 1995 to December 2000. J. Dairy Sci. 85: 295-300.

[16] Pokorny J (2005) Volumetric analysis of oxidized lipids. In Kamal-Eldin A, Pokorny J, editors. Analysis of lipid oxidation. AOCS Publishing.

[17] Simon P (2006) Induction periods theory and applications. J. Therm. Anal. Calorim. 84: 263-270.

[18] Fasina OO, Craig-Schmidt M, Colley Z, Hallman H (2008) Predicting melting characteristics of vegetable oils from fatty acid composition. LWT – Food Sci. Technol. 41: 1501-1505.

[19] Wirkowska M, Ostrowska-Ligeza E, Gorska A, Koczon P (2012) Thermal properties of fats extracted from powdered baby formulas. J. Therm. Anal. Calorim. 110:137-143

[20] Kousksou T, Jamil A, Zeraouli Y, Dumas JP (2006) DSC study and computer modelling of the melting process in ice slurry. Thermochim. Acta. 448: 123-129.

[21] Castello ML, Dweck J, Aranda DAG (2011) Kinetic study of thermal processing of glycerol by thermogravimetry. J. Therm. Anal. Calorim. 105:737-746

[22] Roy S, Riga AT, Alexander KS (2002) Experimental design aids the development of differential scanning calorimetry standard test procedure for pharmaceuticals. Thermochim. Acta. 392-393:399-404.

[23] ISO 11357-6:2008 (E) Plastics – differential scanning calorimetry (DSC) – Part 6: Determination of oxidation induction time (isothermal OIT) and oxidation induction temperature (dynamic OIT).

[24] Tsang W, Walker JA (1985) Characterization of RDF properties through high pressure differential scanning calorimetry. Resour Conserv. 9:355-365.

[25] Adhvaryu A, Perez JM, Singh ID (1999) Application of quantitative NMR spectroscopy to oxidation kinetics of base oils using pressurized differential scanning calorimetry technique. Energ Fuels. 13:493-498.

[26] Butrow AB, Seyler RJ (2003) Vapor pressure by DSC: extending ASTM E 1782 below 5 kPa. Thermochim. Acta. 402:145-152.

[27] Kousksou T, Jamil A, El-Omari K, Zeraouli Y, Le Guer Y (2011) Effect of heating rate and simple geometry on the apparent specific heat capacity: DSC applications. Themochim. Acta. 519:59-64.

[28] Litwinienko G, Daniluk A, Kasprzycka-Guttman T (2000) Study on autoxidation kinetics of fats by differential scanning calorimetry. 1. Saturated C_{12}-C_{18} fatty acids and their esters. Ind. Eng. Chem. Res. 39:1.12.

[29] Vyazovkin S, Wight CA (1997) Kinetics in solids. Annu. Rev. Phys. Chem. 48:125-149.

[30] Bouzidi L, Boodhoo M, Humphrey KL, Narine SS (2005) Use of the first and second derivatives to accurately determine key parameters of DSC thermographs in lipid crystallization studies. Thermochim. Acta. 439:94-102.

[31] Šimon P (2007) The single-step approximation attributes, strong and weak sides. J. Therm. Anal. Calorim. 88:709-715.

[32] Vyazovkin S, Wight CA (1999) Model-free and model-fitting approaches to kinetic analysis of isothermal and non-isothermal data. Thermochim. Acta. 341:53-68.

[33] Šimon P, Kolman L, Niklova I, Schmidt S (2000) Analysis of the induction period of oxidation of edible oils by differential scanning calorimetry. J. Am. Oil Chem. Soc. 77:639-642.

[34] Šimon P, Kolman L (2001) DSC study of oxidation induction periods. J. Therm. Anal. Calorim. 64:813-820.

[35] Schmid M, Ritter A, Affolter S (2006) Determination of oxidation induction time and temperature by DSC. J. Therm. Anal. Calorim. 2:367-371.

[36] Šimon P, Thomas PS, Okuliar J, Ray AS (2003) An incremental integral isoconversional method determination of activation parameters. J. Therm. Anal. Calorim. 72:867-874.

[37] Šimon P, Hynek D, Malikova M, Cibulkova Z (2008) Extrapolation of accelerated thermooxidative tests to lower temperatures applying non-Arrhenius temperature functions. J. Them. Anal. Calorim. 3:817-821.

[38] Šimon P (2009) Material stability predictions applying a new non-Arrhenian temperature function. J. Therm. Anal. Calorim. 2:391-396.

[39] Schmid M, Affolter S (2003) Interlaboratory tests on polymers by differential scanning calorimetry (DSC): determination and comparison of oxidation induction time (OIT) and oxidation induction temperature (OIT*). Polym. Test. 22:419-428.

[40] Cross CK (1970) Oil stability: A DSC alternative for the active oxygen method. J. Amer. Oil Chem. Soc. 47:229-230.

[41] Yamazaki M, Nagao A, Komomiya K (1980) High pressure differential thermal analysis of fatty acid methyl esters and triglycerides. J. Amer. Oil Chem. Soc. 57:59-60.

[42] Raemy A, Froelicher I, Loeliger J (1987) Oxidation of lipids studied by isothermal heat flux calorimetry. Thermochim. Acta. 114:159-164.

[43] Arain S, Sherazi STH, Bhanger MI, Talpur FN, Mahesar SA (2009) Oxidative stability assessment of *Bauhinia purpurea* seed oil in comparison to tow conventional vegetable oils by differential scanning calorimetry and Rancimat methods. Thermochim. Acta. 484:1-3.

[44] Pardauil JJR, Souza LKC, Molfetta F, Zamian JR, Filho GNR, da Costa CEF (2011) Determination of the oxidative stability by DSC of vegetable oils from the Amazonian area. Bioresource Technol. 102:5873-5877.

[45] Gromodzka J, Wardencki W, Pawlowicz R, Muszyn G (2010) Photoinduced and thermal oxidation of rapeseed and sunflower oils. Eur. J. Lipid Sci. Tech. 112:1229-1235.

[46] Velasco J, Adersen ML, Skibsted LH (2004) Evaluation of oxidative stability of vegetable oils by monitoring the tendency to radical formation. A comparison of electron spin resonance spectroscopy with the Rancimat method and differential scanning calorimetry. Food Chem. 85:623-632.

[47] Kowalski B (1989) Determination of oxidative stability of edible vegetable oils by pressure differential scanning calorimetry. Thermochim. Acta. 156:347-358.

[48] Kowalski B, Ratusz K, Miciula A, Krygier K (1997) Monitoring of rapeseed oil autoxidation with a pressure differential scanning calorimeter. Themochim. Acta. 307:177-121.

[49] Rudnik E, Szczucinska A, Gwardiak H, Szulc A, Winiarska A (2001) Comparative studies of oxidative stability of linseed oil. Thermochim. Acta. 370:135-140.

[50] Kowaslki B, Ratusz K, Kowalska D, Bekas W (2004) Determination of the oxidative stability of vegetable oils by differential scanning calorimetry and Rancimat measurements. Eur. J. Lipid Sci. Tech. 106:165-169.

[51] Tan CP, Che Man YB, Selamat J, Yusoff MSA (2001) Application of Arrhenius kinetics to evaluate oxidative stability in vegetable oils by isothermal differential scanning calorimetry. J. Am. Oil Chem. Soc. 78:1133-1138.

[52] Tan CP, Che Man YB, Selamat J, Yusoff MSA (2002) Comparative studies of oxidative stability of edible oils by differential scanning calorimetry and oxidative stability index methods. Food Chem. 76:385-389.

[53] Litwinienko G, Kasprzycka-Guttman, Jarosz-Jarszewska M (1995) Dynamic and isothermal DSC investigation of the kinetic of thermooxidative decomposition of some edible oils. J. Therm. Anal. 45:741-750.

[54] Litwinienki G, Daniluka A, Kasprzycka-guttman T (1999) A differential scanning calorimetry study on the oxidation of C_{12}-C_{18} saturated fatty acids and their esters. J. Am. Oil Chem. Soc. 76:655-657.

[55] Cifti ON, Kowalski B, Gogus F, Fadiloglu S (2009) Effect of the addition of a cocoa butter-like fat enzymatically produced from olive pomace oil on the oxidative stability of cocoa butter. J. Food Sci. 4:E184-E190.

[56] Litwinienko G (2001) Autoxidation of unsaturated fatty acids and their esters. J. Therm. Anal. Calorim. 65:639-646.

[57] Litwinienko G, Kasprzycka-Guttman T, Studzinski M (1997) Effects of selected phenol derivatives on the autoxidation of linolenic acid investigated by DSC non-isothermal methods. Thermochim. Acta. 307:97-106.

[58] Ulkowski M, Musialik M, Litwinienko G (2005) Use of differential scanning calorimetry to study lipid oxidation. 1. Oxidative stability of lecithin and linolenic acid. J. Agric. Food. Chem. 53:9073-9077.

[59] Litwinienko G, Kasprzycka-Guttman T (1998) A DSC study on thermoxidation kinetics of mustard oil. Thermochim. Acta. 319:185-191.

[60] Ostrowska-Ligeza E, Bekas W, Kowalska D, Lobacz M, Wroniak M, Kowalski B (2010) Kinetics of commercial olive oil oxidation: Dynamic differential scanning calorimetry and Rancimat studies. Eur. J. Lipid Sci. Tech. 112:268-274.

[61] Jimenez-Marquez J, Beltran-Maza G (2003) Aplicación de la calorimetría diferencial de barrido (CDB) en la caracterización del aceite de oliva virgen. Grasas Aceites. 54:403-409.

[62] Vittadini E, Lee JH, Frega NG, Min DB, Vodovotz Y (2003) DSC determination of thermally oxidized olive oil. J. Am. Oil Chem. Soc. 8:533-537.

[63] Thurgood J, Ward R, Martini S (2007) Oxidation kinetics of soybean oil/anhydrous milk fat blends: A differential scanning calorimetry study. Food Res. Int. 40:1030-1037.

[64] Litwinienko G, Kasprzycka-Guttman T (2000) Study on the autoxidation kinetics of fat components by differential scanning calorimetry. 2. Unsaturated fatty acids and their esters. Ind. Eng. Chem. Res. 39:13-17.

[65] Polavka J, Paligova J, Cvengroš J, Šimon P (2005) Oxidation stability of methyl esters studied by differential thermal analysis and Rancimat. J. Am. Oil Chem. Soc. 82:519-524.

[66] Quinchia LA, Delgado MA, Valencia C, Franco JM, Gallegos C (2011) Natural and synthetic antioxidant additives for improving the performance of new biolubricant formulations. J. Agr. Food Chem. 59:12917-12924.

[67] Kowalski B (1991) Thermal-oxidative decomposition of edible oils and fats. DSC stidies. Thermochim. Acta. 184:49-57.

[68] Musialik M, Litwinienko G (2007) DSC study of linolenic acid autoxidation inhibited by BHT, dehydrozingerone and olivetol. J. Therm. Anal. Calorim. 88:781-785.

[69] Agrawal RK (1989) The compensation effect – a fact or a fiction. J. Therm. Anal. 35:909-917.

[70] Agrawal RK (1986) On the compensation effect. J. Therm. Anal. 31:73-86.

[71] Lopez C, Lavigne F, Lesieur P, Bourgaux C, Ollivon M (2001) Thermal and structural behavior of milk fat. 1. Unstable species of anhydrous milk fat. J. Dairy Sci. 84:756-766.

[72] Jensen RG (2002) The composition of bovine milk lipids: January 1995 to December 2000. J. Dairy Sci. 85:295-300.

[73] Lock AL, Bauman DE (2004) Modifying milk fat composition of dairy cows to enhance fatty acids beneficial to human health. Lipids. 39:1197-1206.

[74] Bell JA, Griinari JM, Kennelly JJ (2006) Effect of safflower oil, flaxseed oil, monensin, and vitamin E on concentration of conjugated linoleic acid in bovine milk fat. J. Dairy Sci. 89:733-748.

[75] Martinez-Monteagudo SI, Khan M, Saldaña MDA, Temelli F (2011) Obtaining a milk fraction rich in conjugated linoleic acid. Abstract 45, In 24th Canadian Conference on Fats and Oils, September 26-27th, Edmonton, AB, Canada.

[76] Prado GHC, Khan M, Saldaña MDA, Temelli F (2012) Enzymatic hydrolysis of conjugated linoleic acid-enriched anhydrous milk fat in supercritical carbon dioxide. J. Supercrit. Fluids. 66:198-206.

[77] Campbell W, Drake MA, Larick DK (2003) The impact of fortification with conjugated linoleic acid (CLA) on the quality of fluid milk. J. Dairy Sci. 86:43-51.

[78] Mathys A, Reineke K, Heinz V, Knorr D (2009) High pressure thermal sterilization – development and application of temperature controlled spore inactivation studies. High Pressure Res. 29:3-7.

[79] Knoerzer K, Buckow R, Versteeg C (2010) Adiabatic compression heating coefficients for high-pressure processing – a study of some insulating polymer materials. J. Food Eng. 98:110-119.

[80] Wilson DR, Dabrowski L, Stringer S, Moezelaar R, Brocklehurst TF (2008) High pressure in combination with elevated temperature as a method for the sterilization of food. Trends Food Sci. Tech. 19:289-299.

[81] Oley I, Verlinde P, Hendrickx M, Van Loey A (2006) Temperature and pressure stability of L-ascorbic acid and/or [6s] 5-methyltetrahydrofolic acid: A kinetic study. Eur. Food Res. Tech. 223:71-77.

[82] Butz P, Serfert Y, Fernandez-Garcia A, Dieterich S, Lindauer R, Bognar A, Tauscher B (2004) Influence of high-pressure treatment at 25°C and 80°C on folates in orange juice and model media. J. Food Sci. 69:117-121.

[83] Matser AA, Krebbers B, van den Berg RW, Bartels PV (2004) Advantages of high pressure sterilisation on quality of food products. Trends Food Sci. Tech. 15:79-85.

[84] Verbeyst L, Oey I, Van der Plancken I, Hendrickx M, Van Loey A (2010) Kinetic study on the thermal and pressure degradation of anthocyanins in strawberries. Food Chem. 123:269-274.

[85] Martinez-Monteagudo SI, Saldaña MDA, Torres JA, Kennelly JJ. (2012). Effect of pressure-assisted thermal sterilization on conjugated linoleic acid (CLA) content in CLA-enriched milk, Innov Food Sci Emerg Technol, doi 10.1016/j.ifset.2012.07.004.

[86] Martinez-Monteagudo SI, Saldaña MDA (2011) Changes in conjugated linoleic acid (CLA) and antiradical ability of anhydrous milk fat treated with pressure assisted thermal sterilization. Abstract 43, In 24th Canadian Conference on Fats and Oils, September 26-27th, Edmonton, AB, Canada.

[87] Fagali N, Catala A (2008) Antioxidant activity of conjugated acid isomers, linoleic acid and its methyl ester determined by photoemission and DPPH techniques. Biophys. Chem. 137:56-62.

Chapter 10

Differential Scanning Calorimetry Studies of Phospholipid Membranes: The Interdigitated Gel Phase

Eric A. Smith and Phoebe K. Dea

Additional information is available at the end of the chapter

1. Introduction

DSC is a versatile technique and has been used for decades to study hydrated phospholipid membranes [1-4]. It can even be used to analyze whole cell samples [5]. For pure lipids, DSC can accurately determine the phase transition temperatures and the associated enthalpies. As a consequence, how the chemical structure of lipids translates into thermodynamic properties can be systematically studied. In addition to determining the physical properties of pure lipids, the miscibility and phase behavior of lipid mixtures can be determined.

The detailed review of the interdigitated phase written by Slater and Huang in 1988 provides an excellent outline of the properties of the interdigitated phase and the relevant analytical techniques [6]. Furthermore, the meticulous studies of Koynova and Caffrey describe how systematic changes in lipid chemistry can affect their phase behavior [7-9]. Lipids with asymmetrical acyl chains that form either mixed- or partially-interdigitated phases have also been thoroughly investigated [7,10-12]. This review focuses on the interdigitated phase of fully hydrated phospholipids with hydrocarbon chain lengths of equal size. We pay special attention to recently discovered interdigitated systems and the chemicals that can induce or inhibit lipid interdigitation.

For simplicity, we have centered our review around the extensively studied lipid, 1,2-dipalmitoyl-*sn*-glycero-3-phosphocholine (DPPC). DPPC is naturally occurring and has thermodynamic phase behavior that is typical for saturated phosphatidylcholines (PCs) [7]. Although DPPC does not spontaneously interdigitate when hydrated, it can be reliably transformed into the fully interdigitated gel phase (Tables 1 and 2). Alterations in the lipid hydrocarbon chains (Figure 1) and the lipid head group (Figure 2) substantially affect spontaneous interdigitation (Figure 3). The predisposition for interdigitation is a finely tuned balance of interactions among hydrocarbon chains, between polar head groups, and at the interfacial area with the aqueous phase.

DPPC DHPC F-DPPC DOPC DEPC

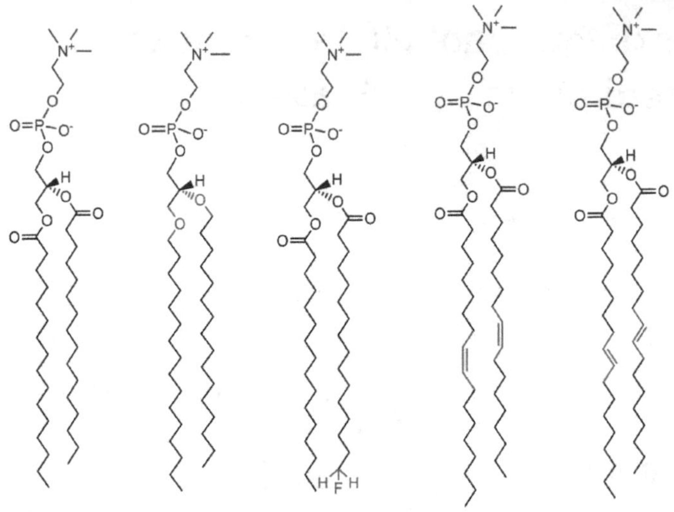

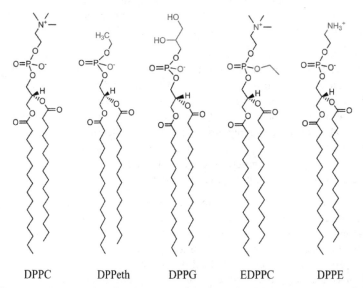

Figure 1. Representative examples of fatty acid modifications of DPPC. The segments in red identify modifications to the structure of DPPC.

DPPC DPPeth DPPG EDPPC DPPE

Figure 2. Representative examples of head group modifications of DPPC. The segments in red identify modifications to the structure of DPPC. The segments in blue demonstrate the resulting change in charge.

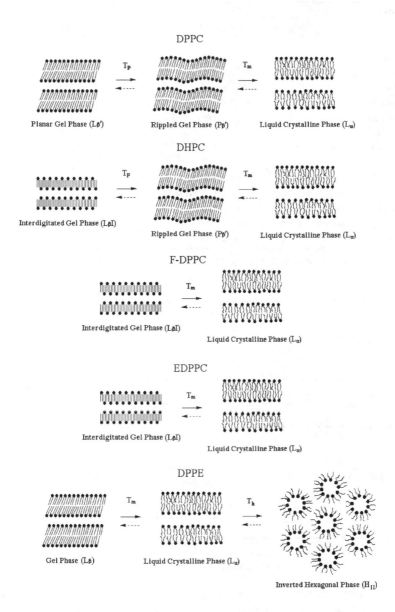

Figure 3. Phase transitions of representative lipids. Solid arrows indicate heating transitions and dotted arrows indicate cooling transitions. Note: subgel phase transitions are not shown.

DPPC/alcohol	Threshold concentration (M)	References
Methanol	2.75 ± 0.35	[13]
Ethanol	1.10 ± 0.10	[13]
1-propanol	0.39 ± 0.03	[14]
2-propanol	0.52 ± 0.03	[14]
1-butanol	0.16 ± 0.02	[15]
Isobutanol	0.17 ± 0.02	[15]
sec-butanol	0.22 ± 0.02	[15]
tert-butanol	0.27 ± 0.02	[15]
1-pentanol	0.07 ± 0.01	[16]
2-pentanol	0.10 ± 0.01	[16]
3-pentanol	0.11 ± 0.01	[16]
3-methyl-2-butanol	0.10 ± 0.01	[16]
2-methyl-1-butanol	0.08 ± 0.01	[16]
3-methyl-1-butanol	0.08 ± 0.01	[16]
2-methyl-2-butanol	0.13 ± 0.01	[16]
neopentanol	0.08 ± 0.01	[16]

Table 1. Threshold concentrations for some alcohol-induced interdigitation.

	Chemicals	References
	bupivacaine	[17]
	dibucaine	[17]
Anesthetics	lidocaine	[17]
	procaine	[17]
	tetracaine	[17-19]
	labdanes	[20]
Drugs	chlorpromazine	[19]
	valsartan	[21]
	glycerol	[22-24]
	ethylene glycol	[25]
Organic solvents	acetone	[26,27]
	acetonitrile	[27,28]
	propionaldehyde	[27]
	petrahydrofuran	[27]
Salts	KSCN	[29,30]
Pressure	N/A	[31-34]

Table 2. Induced interdigitation of PC membranes.

There are multiple types of interdigitation (Figure 4). The type of interdigitation that forms is heavily dependent on the structure and symmetry of the hydrocarbon chains [6]. Interdigitated lipid systems can further be separated into two classes: spontaneous and induced. We use "spontaneous" to describe lipids that self-assemble into the interdigitated gel phase when fully hydrated under typical preparation procedures and at ambient pressure. Some notable recent examples are highlighted, such as cationic lipids and lipids with monofluorinated acyl chains. Whether or not a particular lipid spontaneously interdigitates is determined by the balance of properties that favor and disfavor interdigitation. Lipids often have conflicting characteristics regarding the ability to form the interdigitated phase. Consequently, there is no simple formula for determining which lipids will spontaneously interdigitate without relying on experimental data.

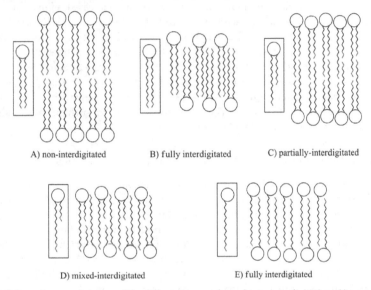

A) non-interdigitated   B) fully interdigitated   C) partially-interdigitated

D) mixed-interdigitated   E) fully interdigitated

Figure 4. Schematic representation of the different types of membrane interdigitation: A) symmetrical lipid, non-interdigitated; B) symmetrical lipid, fully interdigitated; C) asymmetrical lipid, partially interdigitated; D) highly asymmetrical lipid, mixed-interdigitated; E) lysolipid, fully interdigitated. For clarity, the terminal ends of the hydrocarbon chains are labeled in red.

As can be seen in Figure 4, the structural difference between the interdigitated and non-interdigitated phases can be substantial. In the non-interdigitated membrane, both ends of the hydrocarbon chains meet in the membrane midplane (Figure 4A). Two well-defined leaflets are formed and there is a thick hydrophobic core. In the fully-interdigitated membrane, the thickness of the interdigitated phase is greatly reduced and there is the loss of the membrane midplane. There is an increase in the spacing between the polar lipid head groups and the ends of the lipid hydrocarbon chains become more exposed to the aqueous interface [6]. The difference is most dramatic in the fully interdigitated phase compared to the non-interdigitated membrane (Figure 4A and 4B). In the partially-interdigitated system, the longer chain extends to the other side of the membrane and aligns with the apposing shorter chain (Figure 4C). In the mixed-interdigitated membrane, the short hydrocarbon chains line up with each other and the full-length chain extends to the other side of the membrane (Figure 4D). Lyso lipids also form a fully interdigitated structure (Figure 4E) [35].

2.

2.1. Thermodynamics of phosphatidylcholine membranes

PCs are common in mammalian membranes and have well known phase transitions [7]. The most ubiquitous of these is the gel-to-liquid crystalline transition, often referred to as the melting or main transition. This transition is relatively rapid and is highly reversible [36]. It is characterized by the co-operative melting of the hydrocarbon chains and a high enthalpy DSC peak [2]. The liquid crystalline (L_α) phase has an increased number of gauche conformers and a large increase in membrane fluidity and disorder [2,37]. The pre-transition from the planar gel ($L_{\beta'}$) to the rippled gel phase ($P_{\beta'}$) has a low enthalpy and is sensitive to sample preparation and the presence of impurities [7,36]. It is also more sensitive to the scan rate, with lower scan rates resulting in lower T_P temperatures [3]. Some PCs also have subgel phases. The subgel transition is slow and is dependent on sample preparation, especially incubation temperature and time [36]. All of the above phases are strongly affected by changes in lipid structure. This review will show that such alterations also have a profound effect on the interdigitated phase.

2.2. Chemically-induced interdigitation of phosphatidylcholines

The most widely studied chemical inducer of interdigitation is ethanol. In non-interdigitated phospholipid membranes, ethanol tends to adsorb to the head groups, especially the region near the hydrocarbon chains [38,39]. In particular, the carbonyl groups of the glycerol backbone of phospholipids are thought to be the favored hydrogen bonding sites for ethanol [40]. Ethanol displaces water when it adsorbs to the head group, which increases the head group volume and decreases the order of the hydrocarbon chains [13,41,42]. The increase in head group volume leads to increased chain tilting and creates energetically unfavorable voids in the hydrocarbon region of non-interdigitated membranes, encouraging the creation of the $L_\beta I$ phase at high concentrations [13,39,42-44]. Once the $L_\beta I$ phase is formed, ethanol can bind to the exposed hydrocarbon chains, replacing the unfavorable interaction of the acyl chains with water [45]. Also, it is typical for alcohols to increase the main transition enthalpy above the threshold concentration for interdigitation [15,16].

There are three main characteristics of the chemically-induced interdigitated phase in the DSC thermograms of saturated PCs: the presence of biphasic phase behavior, an increase in T_m hysteresis, and the suppression of the pre-transition. The combination of these can be used to determine the threshold concentration for interdigitation.

The "biphasic effect" indicates two independent interactions within different concentration ranges [46,47]. The biphasic effect is most strongly characterized by an initial decrease in the T_m, but an increase in or stabilization of the T_m once the $L_\beta I$ phase is formed. The first interaction lies below the threshold concentration. Here, ethanol preferentially partitions into the liquid crystalline phase, lowering the phase transition temperature [47]. The secondary interaction above the threshold concentration stabilizes the interdigitated gel phase. The main transition co-operativity (sharpness of transition peak) can also be enhanced above the threshold concentration [47].

The shape and magnitude of the T_m biphasic effect is also dependent on the alcohol isomer used as shown in Figure 5 [15,16]. For example, *tert*-butanol is an effective inducer of interdigitation and has a pronounced biphasic effect [15]. Other alcohols, such as the pentanol isomers have more "stunted" biphasic behavior [16]. While a difference in trend can still be observed above and below the threshold concentration, the distinction is less pronounced.

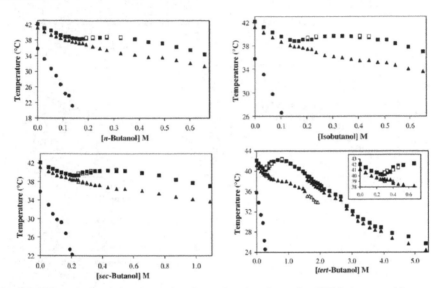

Figure 5. Effects of *n*-butanol, isobutanol, *sec*-butanol, and *tert*-butanol on DPPC phase transition temperatures. (■, heating scan main peak; □, heating scan shoulder peak; ▲, cooling scan main peak; Δ, cooling scan shoulder peak; •, pre-transition peak). Reprinted from Biophys. Chem., 128, Reeves MD, Schawel AK, Wang W, Dea PK, Effects of butanol isomers on dipalmitoylphosphatidylcholine bilayer membranes, Pages No. 13-18, Copyright (2007), with permission from Elsevier [15].

Increasing the alcohol content well above the threshold concentration lowers the T_m [15,16]. At these concentrations, additional alcohol destabilizes the $L_\beta I$ phase relative to the L_α phase. The membrane bilayer structure can also break down for alcohols that are highly soluble in water [15]. For example, above 2.00 M *tert*-butanol in DPPC, the main transition hysteresis is absent (Figure 5). Additionally, the heating main transition DSC peaks above 2.00 M *tert*-butanol become increasingly broad. Changes in the ^{31}P-NMR spectra at high concentrations confirm the loss of lamellar structure [15].

The biphasic behavior is also reflected in the increase in the main transition enthalpy as the alcohol concentration increases [15,16]. Often, the rate of change in the transition enthalpy above and below the threshold concentration is different (Figure 6). This effect also depends on the alcohol chain length and isomer used. For instance, this difference is clear with *n*-butanol but not *tert*-butanol [15].

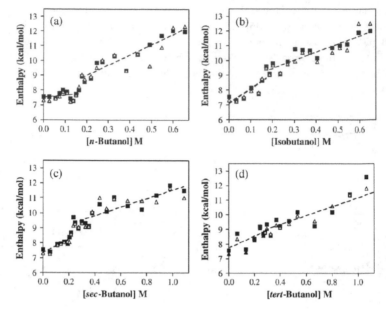

Figure 6. Effects of (a) *n*-butanol, (b) isobutanol, (c) *sec*-butanol and (d) *tert*-butanol on DPPC main transition enthalpies (■, heating scan main transition enthalpy; Δ, cooling scan main transition enthalpy). Reprinted from Biophys. Chem., 128, Reeves MD, Schawel AK, Wang W, Dea PK, Effects of butanol isomers on dipalmitoylphosphatidylcholine bilayer membranes, Pages No. 13-18, Copyright (2007), with permission from Elsevier [15].

A property that accompanies the biphasic effect is the emergence of hysteresis in the main transition [15,16,48,49]. The hysteresis as it relates to DSC is defined as the difference in the transition temperature between heating and cooling scans. This corresponds to the reversibility and kinetics of the transition. The return to the interdigitated phase with a decrease in temperature is a slow process and is therefore less reversible [48,51]. For systems that do not interdigitate, such as phosphatidylethanolamine (PE) lipids, the addition of alcohol does not affect the transition hysteresis [48].

The disappearance of the pre-transition is another consistent property of alcohol-induced interdigitation of saturated PCs. The decrease in the T_P follows a well defined trend below the threshold concentration until it is finally abolished (Figure 5). The rate at which the T_P is depressed depends on the efficacy of the chemical inducer.

By comparing the threshold concentrations of different chemicals, they can be ranked on their effectiveness at inducing the interdigitated phase. For instance, the threshold concentrations for alcohol-induced interdigitation systematically decreases as the lipid hydrocarbon chain length increases (Table 1). The isomers with the most solubility in water are the least effective at inducing interdigitation, as shown by the increase in threshold concentrations [15]. Additionally, the more soluble an isomer is in water, the less effectively it depresses both the temperature of the pre-transition and the main transition prior to the threshold concentration [15].

The ether-linked analogue of DPPC, 1,2-di-O-hexadecyl-sn-glycero-3-phosphocholine (DHPC), is also a useful model membrane for studying interdigitation. DHPC goes through a low temperature pre-transition from the $L_\beta I$ phase to the non-interdigitated rippled gel phase $P_{\beta'}$ (Figure 3) [52,53]. Therefore, chemicals that stabilize the $L_\beta I$ phase increase the T_P until it merges with the main transition into the L_α phase [49]. This process occurs at lower concentrations for more effective inducers of interdigitation.

2.3. Pressure-induced interdigitation

It is well established that the application of hydrostatic pressure favors interdigitation in a multitude of lipid systems (Tables 2 and 3). As hydrostatic pressure is applied, the intermolecular distance between adjacent lipids is reduced and molecular packing becomes denser [34]. By changing the packing structure of the membrane, interdigitation can relieve the stress caused by the increased steric hindrance.

Pressure-induced interdigitation is dependent on lipid hydrocarbon chain length and the chemical structure, much like chemically-induced interdigitation. Ether- and ester-linked lipids with longer chains require less pressure to interdigitate [51,54]. Under high temperature and pressure conditions, ester-linked lipids behave similarly to the equivalent ether-linked lipids at normal pressure [51]. Pressure-induced interdigitation is not universal, however. As with chemically-induced interdigitation, certain lipids do not interdigitate even under high pressure [34,55,56].

2.4. Spontaneous interdigitation in ether-linked lipids and 1,3-DPPC

The type of bond that connects the hydrocarbon chain to the lipid head group also affects the thermodynamic properties. Switching either or both of the ester bonds of DPPC with ether linkages results in a small increase in the T_m (<4 °C) and enthalpy (<1 kcal/mol) [57]. A single ether linkage can be sufficient to allow the formation of the interdigitated gel phase [57,58]. Furthermore, in ether lipids that spontaneously interdigitate, the interdigitated phase is stable up to higher temperatures as the chain length increases (Figure 7) [51]. There is an increased amount of head group repulsion in DHPC, which favors the interdigitated phase [51,59,60]. Conversely, the stronger interactions in the head groups of ester lipids hinder interdigitation [34,51]. DHPC is an especially useful lipid for studying the interdigitated phase because its transition from the interdigitated gel to non-interdigitated ripple gel phase is highly sensitive to its environment.

The similarity of DHPC to DPPC also allows for the comparison between ether- and ester-linked lipids. It is consistently easier to interdigitate ether-linked lipids whether through chemical means [17,48,49] or by the application of pressure [51,61]. Furthermore, the ether-linked 1,2-di-O-hexadecyl-sn-glycero-3-phosphoethanolamine (DHPE) demonstrates the result of competing influences on the interdigitated gel phase. In DHPE, the ether linkages favor interdigitation, but the PE head group is more strongly opposed to interdigitation [62]. Therefore, while DHPC spontaneously interdigitates, the PE head group of DHPE prevents interdigitation.

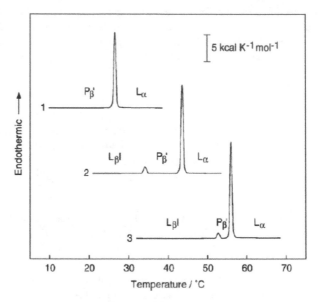

Figure 7. DSC heating thermograms of ether-linked PC bilayer membranes: (1) O-14:0-PC, (2) O-16:0-PC, (3) O-18:0-PC Reprinted from Biochim. Biophys. Acta, 1768, Matsuki H, Miyazaki E, Sakano F, Tamai N, Kaneshina S, Thermotropic and barotropic phase transitions in bilayer membranes of ether-linked phospholipids with varying alkyl chain lengths, Pages No. 479-489, Copyright (2007), with permission from Elsevier [51].

While the majority of PC lipid studies use lipids with the hydrocarbon chains on the sn-1 and sn-2 positions, there are some examples of experiments using synthetic lipids with the chains located at sn-1 and sn-3. One intriguing example is the positional isomer of DPPC, 1,3-dipalmitoyl-sn-glycero-2-phosphocholine (1,3-DPPC or β-DPPC), which has unique properties including the ability to spontaneously interdigitate [63-65]. However, the phase diagram is different from the ether-linked lipids that spontaneously interdigitate [51,66]. At lower temperatures, 1,3-DPPC exists in a non-interdigitated "crystalline" bilayer phase termed (L_c). At higher temperatures, but below the T_m, 1,3-DPPC can form a fully interdigitated structure [63]. The ability to interdigitate may be due to greater head group repulsion resulting from a different phosphocholine tilt or conformation relative to the glycerol backbone [63-65]. As with most interdigitated systems, 1,3-DPPC converts into a non-interdigitated structure during the heating transition into the L_α phase. The cooling transition from the L_α phase into the interdigitated phase has considerable hysteresis [63].

2.5. The monofluorinated analogue of DPPC: F-DPPC

The monofluorinated analogue of DPPC, 1-palmitoyl-2-(16-fluoropalmitoyl)sn-glycero-3-phosphocholine (F-DPPC), spontaneously forms the $L_\beta I$ phase below the main transition temperature (T_m) [67-69]. The main transition temperature of F-DPPC also occurs at a higher temperature (~50 °C) and with a higher transition enthalpy (9.8 kcal/mol) compared to DPPC [67]. The endothermic peak of F-DPPC is also broader than DPPC. The transition can be split into two overlapping peaks, with the peak centered at 50.6 °C accounting for 36% of the area and the peak at 52.0 °C accounting for 64% of the area [67]. The lower transition peak component does not correspond to a change in the hydrocarbon chains as detected by

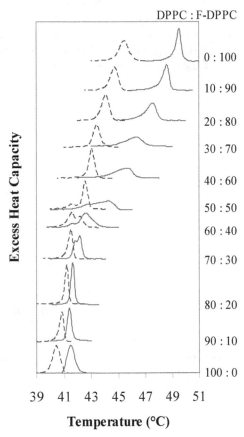

Figure 8. Heating (red lines) and cooling (dashed blue lines) DSC thermograms of the T_m of the DPPC/F-DPPC system are shown. The cooling scans have been inverted to allow comparison with the heating thermograms. For clarity, the thermograms are also offset vertically. Reprinted from Biophys. Chem., 147, Smith E.A., van Gorkum C.M., Dea P.K., Properties of phosphatidylcholine in the presence of its monofluorinated analogue, Pages No. 20-27, Copyright (2010), with permission from Elsevier [69].

FTIR spectroscopy. It is possible that this relates to a conversion from interdigitated to non-interdigitated gel right before the transition into the liquid crystalline phase [67]. The main transition is also characterized by a large main transition hysteresis (Figures 8 and 9) [67,69]. Additionally, the $L_\beta I$ phase has high conformational order and tight lipid packing [68].

It appears that the fluorine must be located on the terminal hydrocarbon chain to have a dramatic effect on interdigitation. When the fluorine substitution is not located on the terminal carbon, DSC data reveal that the physical properties are only modestly changed and they are largely miscible with the non-fluorinated parent lipid [70]. Lipids with more fluorine, such as when the 13-16 carbons are perfluorinated, do not spontaneously interdigitate either [71,72]. Therefore, it is the interaction of the polar terminal C-F bond with the aqueous interface that encourages interdigitation [67]. The large dipole moment is the most likely culprit for stabilizing the interdigitated phase by reducing the unfavorable exposure of the hydrophobic acyl chains to water. However, the slightly larger van der Waals radius and the possibility of weak hydrogen bonding may also play a role [73-77].

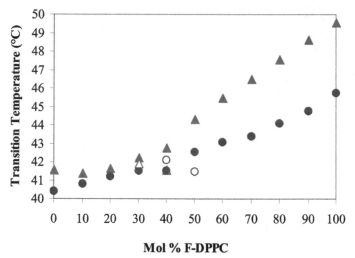

Mol % F-DPPC

Figure 9. The main transition temperature (T_m) of DPPC/F-DPPC mixtures determined by DSC. Heating scans shown by filled red triangles (▲). Cooling scans shown by filled blue circles (•). Shoulder peaks indicated by unfilled triangles for heating scans (△) and unfilled circles for cooling scans (○). Reprinted from Biophys. Chem., 147, Smith EA, van Gorkum CM, Dea PK, Properties of phosphatidylcholine in the presence of its monofluorinated analogue, Pages No. 20-27, Copyright (2010), with permission from Elsevier [69].

2.6. The interdigitated gel phase in anionic lipids

As with PCs, di-saturated long chain phosphatidylglycerols (PGs) have a strong propensity towards interdigitation (Table 3) [78]. The negatively charged PGs are commonly found in microbial membranes [90]. The interaction of some peptides with lipids is heavily dependent on the composition of the membrane [82]. This contributes to the ability of antimicrobial peptides to selectively target microbial membranes [91]. Recently, it was found that DPPG has the ability to form a quasi-interdigitated gel phase with the addition of the human multifunctional peptide LL-37 [81,82]. The antimicrobial peptide peptidyl-glycylleucine-carboxyamide (PGLa) has a similar effect below the main transition temperature of saturated PGs [83]. In these instances, the peptide shields the acyl chains of the interdigitated lipid from the aqueous layer by orienting in the interfacial region below the T_m (Figure 10).

Furthermore, other chemicals such as Tris-HCl induce interdigitation in DPPG by binding between lipids, resulting in the increased area per head group that favors interdigitation [85]. As in zwitterionic lipids, interdigitation relieves head group repulsion in charged lipids by allowing for a larger area per head group [84]. Charge repulsion in DPPG leads to tilted acyl chains in the non-interdigitated bilayer [85]. This is similar to the ethanol-induced interdigitation of DPPC, where the increased head group size increases the tilt in the gel phase and which ultimately results in the interdigitated gel phase [43]. Ethanol further enhances interdigitation in DPPG, most likely by partitioning into the interfacial region and reducing the exposure of the terminal methyl groups to water [84].

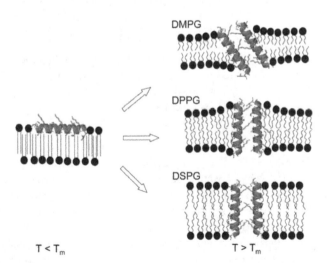

Figure 10. Schematic representation of the peptide PGLa-associated structural changes in PG bilayers. Below the main phase transition ($T < T_m$), the lipids of different hydrocarbon chain lengths exhibit a quasi-interdigitated phase in the presence of PGLa. Reprinted from Biophys. J. 95, Pabst G, Grage SL, Danner-Pongratz S, Jing W, Ulrich AS, Watts A, Lohner K, Hickel A, Membrane thickening by the antimicrobial peptide PGLa, Pages No. 5779-5788, Copyright (2008), with permission from Elsevier [83].

Inducer	References
polymyxin B	[24,79,80]
peptide LL-37	[81,82]
peptide PGLa	[83]
myelin basic protein	[80]
Tris HCl	[79,84,85]
choline and acetylcholine	[86]
atropine	[87,88]
anisodamine	[89]
Pressure	[50]

Table 3. Induced interdigitation of PG membranes

When ethanol substitutes for water in the transphosphatidylation reaction catalyzed by phospholipase D, phosphatidylethanols (Peth) are formed [84,92]. Peth lipids are unique because they have a small anionic lipid headgroup (Figure 2). These lipids are biologically relevant since Peths accumulate in membranes of animal models of alcoholism [93]. Like DPPG, DPPeth can be chemically induced to interdigitate with Tris-HCl and the interdigitated phase is stabilized with the addition of ethanol [84].

2.7. The interdigitated gel phase in cationic lipids

Cationic lipids with modified head groups can spontaneously form interdigitated gel phases below the main transition. One recent example is the positively charged lysyl-DPPG, which is DPPG with a lysine moiety attached. Lysyl-DPPC forms an interdigitated phase primarily due to the large repulsion between head groups [94].

Another modification is the esterification of the phosphate head group, which increases the steric bulk and changes the molecule from zwitterionic to positively charged, allowing interdigitation [95,96]. For example, the P-O-ethyl ester analogue of DPPC, 1,2-dipalmitoyl-sn-glycero-3-ethylphosphocholine (EDPPC or Et-DPPC), is fully interdigitated in the gel phase and has a main transition at 42.5 °C with an enthalpy of 9.6 kcal/mol [95-97]. These values are slightly higher than those for DPPC, which has a T_m around 41.3 °C and a corresponding enthalpy of 8.2 kcal/mol ([7] and references therein). The thermodynamic behavior of these cationic triesters of phosphatidylcholine can be attributed to the net positive charge and the absence of intermolecular hydrogen bonding [98]. Furthermore, the overall polarity of the lipid is decreased, which may decrease the interfacial polarity. This would reduce the energetic cost when the ends of the hydrocarbon chains are exposed to the polar head group and the aqueous phase in the interdigitated phase [96]. The additional ethyl group in the head group may also mitigate the unfavorable exposure of the acyl chains [96]. Infrared spectroscopic data lend some support to this conclusion, since the polar/apolar interfaces of cationic PCs are less polar than the parent PC lipids [96]. These lipids have complex phase diagrams that are dependent on temperature, mechanical agitation, and kinetics (Figure 11) [97].

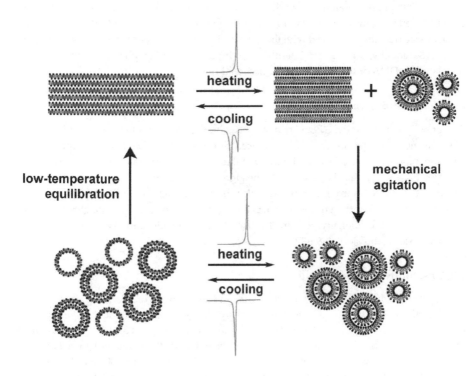

Figure 11. Diagram of the morphological changes in EDPPC dispersions. The equilibrium low temperature arrangement appears to be lamellar sheets, with chain interdigitation. Upon heating, liposomes and lamellar sheets (both non-interdigitated) coexist, whose mixture fully converts into liposomes (apparently the equilibrium liquid crystalline phase arrangement) only after mechanical treatment. Cooling back to the gel phase produces gel-phase liposomes which convert back into lamellar sheets only after prolonged low-temperature exposure. Reprinted from Biochim. Biophys. Acta., 1613, Koynova R, MacDonald RC, Cationic O-ethylphosphatidylcholines and their lipoplexes: phase behavior aspects, structural organization and morphology, Pages No. 39-48, Copyright (2003), with permission from Elsevier [97].

Vesicles formed from cationic triester lipids readily fuse with anionic lipids [98]. This may help explain why lipoplexes made from cationic o-ethylphosphatidylcholines with di-saturated hydrocarbon chains are effective transfection agents [98,99]. The structure and transfection capability of cationic phospholipid-DNA complexes are dependent on preparation conditions and ionic strength [98,100]. These lipids confer multiple advantages: they are non-viral, metabolized by cells, have low toxicity, and closely resemble naturally occurring phospholipids [101].

In some instances, DNA can be sandwiched between interdigitated gel phase lipid sheets into a rectangular columnar two-dimensional superlattice [97,99]. The gel-to-liquid crystalline phase transition results in the contraction of the DNA strand arrays so that the mean charge density is balanced with the increased positive charge of the non-interdigitated lipid. Additionally, in the non-interdigitated liquid crystalline phase, the interlamellar correlation in DNA ordering is no longer observed [99].

2.8. Phosphatidylethanolamines

Phosphatidylethanolamines (PEs) are distinct due to their strong reluctance to interdigitate. PEs are not susceptible to alcohol-induced interdigitation [48] or pressure-induced interdigitation [55]. Even the ether-linked DHPE does not interdigitate with pressure [34,56]. A major reason for this is that the PE headgroup can form hydrogen bonds [34]. PC headgroups interact through a weaker electrostatic attraction between the positively charged quaternary nitrogen and the negatively charged oxygen of a neighboring lipid headgroup [34]. Additionally, the smaller size of the PE headgroup also allows for closer interaction (less repulsion) [2,8,102].

2.9. Unsaturated phospholipids

Unsaturated lipids with common head groups and acyl chains of nearly equal length are strongly disfavored to interdigitate spontaneously. In general, unsaturated lipids are also resistant to both pressure- and chemically-induced interdigitation [44,103,104]. Even under high pressure, unsaturated lipids typically retain the transition from the non-interdigitated lamellar gel phase (L_β) into the liquid crystalline phase (L_α) [104,105]. Pressure does stabilize the L_β phase to the detriment of the L_α phase, but it is not sufficient to induce interdigitation [105].

Furthermore, unsaturated lipids have substantially lower main transition temperatures [7,10]. As interdigitation is highly unfavorable in the liquid crystalline phase, the relevant temperature range of the gel phase where interdigitation is likely to occur is much smaller. The main transition tends to be lowered the most when the double bond is located near the middle of the fatty acid chain [2,10,36]. Although double bonds that are *trans* usually have less influence than *cis* bonds [7,36], no ethanol-induced interdigitation was found in the *trans* lipid 1,2-dielaidoyl-*sn*-glycero-3-phosphocholine (DEPC) [106]. It was postulated that the increased cross-sectional area due to the double bond and the restriction in sliding motions contributes to the lack of interdigitation [106].

The inhibition of interdigitation also applies to lipid mixtures involving unsaturated lipids. In a model membrane of DPPC/DOPC/ergosterol, increasing the unsaturated lipid or sterol component co-operatively hinders the formation of the interdigitated phase [107]. DOPC is known to result in a more disordered and less tilted gel phase and can lead to phase separation at higher concentrations [108]. As a consequence, it was hypothesized that changes in the plasma membrane composition may play a role in the ethanol tolerance of yeast cells during fermentation ([107] and references therein).

There are some exceptions, however. While lipids with double bonds on both chains are particularly unlikely to interdigitate, there are a few examples of interdigitation where only one chain has a double bond. For example, McIntosh et al. tested the ethanol-induced interdigitation of five positional isomers of 1-eicosanoyl-2-eicosenoyl-sn-glycero-3-phosphocholine (C(20):C(20:1Δ^n)PC) with a single cis bond on the sn-2 chain at position $n = 5$, 8, 11, 13 and 17 [109]. Ethanol-induced interdigitation can be induced when the position of the cis bond is at $n= 5$ or 8, but not at $n= 11$, 13, or 17 [109]. In contrast, the fully saturated lipid with the same chain length can easily be interdigitated with a small amount of ethanol [47].

Additionally the cis mono-unsaturated 1-stearoyl,2-oleoyl-phosphatidylcholine (SOPC) can be interdigitated with glycerol [24]. The PG lipid, 1-palmitoyl,2-oleoyl-phosphatidylglycerol (POPG), can also be interdigitated with the addition of polymyxin B [24]. Certain mixtures of unsaturated zwitterionic and charged lipids, such as 1-palmitoyl-2-oleoyl-phosphatidylcholine (POPC) and POPG, can form the interdigitated phase at high concentrations of ethanol and at low hydration [110].

Therefore, it can be concluded that while it is possible to induce the interdigitated gel phase in unsaturated lipids, they are more resistant to interdigitation compared to the equivalent saturated lipid. Additionally, when the interdigitated phase does occur in unsaturated lipids, the phase appears to be less stable and less ordered than in saturated lipids [24].

2.10. Membrane curvature and interdigitation

The curvature of the membrane due to the macromolecular size and shape affects the thermodynamic properties [36]. For instance, on DSC scans, small unilamellar vesicles (SUVs) have lower enthalpic peaks and greater widths compared to multilamellar vesicles (MLVs) [2,3]. SUVs also have more mobility and less order in the hydrocarbon chains [2].

The degree of membrane curvature also affects the ability to interdigitate. Bending in the membrane causes increased steric interference in opposing lipid monolayers [44]. As a consequence, ethanol-induced interdigitation is dependent on curvature, with the more highly curved vesicles requiring more ethanol to interdigitate [111,112]. Sonicated DPPC SUVs are not stable in the presence of ethanol above the threshold concentration for interdigitation [112]. Furthermore, SUVs have a tendency to fuse into large unilamellar vesicles (LUVs), which have properties more similar to MLVs [36,113]. The more planar MLVs allow interdigitated lipids to slide by each other with low steric interference and therefore have the lowest threshold concentrations [44,112].

2.11. Inhibition of the interdigitated gel phase by cholesterol

Just as there are chemicals that induce interdigitation, there are chemicals that inhibit the formation of the $L_\beta I$ phase. Cholesterol is a chemical inhibitor of interdigitation in a wide variety of lipid systems (Table 4).

Interdigitated Lipid System	References
DPPC/EtOH	[114]
DPPC/Pressure	[31]
DPPeth/Tris-HCl	[115]
DPPG/Tris-HCl	[115]
DHPC	[66,116]
EDPPC	[95]
F-DPPC	[117]
3:7 ratio of 16:0 LPC:DPPC	[118]

Table 4. The inhibition of interdigitation by cholesterol

In non-interdigitated membranes, cholesterol increases the fluidity of the gel phase, broadens the main transition, and decreases the main transition enthalpy [119]. Figure 12 demonstrates that these effects are also seen in membranes where cholesterol eliminates the interdigitated phase [115-117]. The amount of cholesterol required to prevent interdigitation is related to the stability of the $L_\beta I$ phase. For DHPC, only ~5 mol% cholesterol is required to eliminate interdigitation [66,116]. However, the amount of cholesterol required to prevent interdigitation is approximately quadrupled for F-DPPC, which exists in the $L_\beta I$ phase around 15 °C higher than DHPC [117]. At high cholesterol concentrations the $L_\beta I$ phase of F-DPPC is replaced by a non-interdigitated liquid-ordered (l_o) phase with properties similar to DPPC/cholesterol [117]. On DSC scans, this effect can be observed by the broadening of the main transition peak and a reduction in the T_m hysteresis (Figure 12). The interdigitated phase of cationic EDPPC is especially resilient in the presence of cholesterol, with interdigitated domains still present at 30 mol% cholesterol [95].

There are multiple reasons why cholesterol-rich membranes disfavor interdigitation. For example, lipid head group crowding is mitigated by cholesterol serving as a spacer between lipids [115]. If cholesterol is placed within an interdigitated membrane, the increased spacing also increases the likelihood that the terminal lipid methyl groups will be exposed at the aqueous interface. Since the interdigitated phase lacks the thick membrane midplane region of non-interdigitated membranes, hydrophobic cholesterol located within the interdigitated phase is more likely to come in contact with water [115]. Furthermore, cholesterol significantly disrupts the lattice structure of gel phase lipids [108,117,120]. Lastly, in the interdigitated phase of highly asymmetrical lyso-lipids, cholesterol can take the place of the missing acyl chain thereby compensating for the size mismatch between the head group and the hydrocarbon chains [118,121].

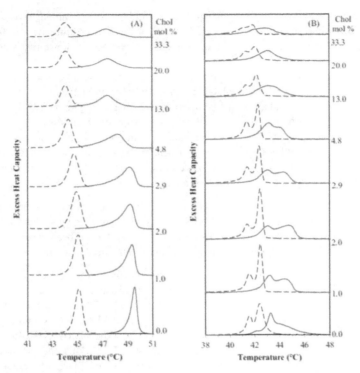

Figure 12. Heating (solid red lines) and cooling (dashed blue lines) DSC thermograms of: (A) F-DPPC and (B) 1:1 F-DPPC/DPPC with various concentrations of cholesterol. The cooling scans have been inverted to allow comparison with the heating peaks. For clarity, the thermograms are also offset vertically. Reprinted from Chem. Phys. Lipids, 165, Smith EA, Wang W, Dea PK, Effects of cholesterol on phospholipid membranes: Inhibition of the interdigitated gel phase of F-DPPC and F-DPPC/DPPC, Pages No. 151-159, Copyright (2012), with permission from Elsevier [117].

2.12. Chemical inhibition of the interdigitated gel phase

In a general sense, solvent inhibitors of interdigitation work in the opposite fashion as chemical inducers. Some researchers have focused on the difference of how kosmotropic and chaotropic solutes interact with lipid membranes [122-124]. Kosmotropes deplete the solution at the interface and increase the interfacial tension whereas chaotropes accumulate in the interface and decrease surface tension [125]. The changes in the structure of water due to these types of chemicals can be attributed to alterations in the hydrogen bonding network of water [126,127]. Kosmotropic substances are classified as water-structure makers, meaning that they stabilize the structure of bulk water. When kosmotropes interact with hydrated lipids, they tend to reduce the interfacial area and inhibit interdigitation [122]. Chaotropic chemicals are classified as water-structure breakers and increase the surface area per lipid, favoring interdigitation [122,124].

The differences between chemicals that induce or inhibit interdigitation have also been illustrated according to the interaction free energy of the lipid membrane interface with solvents ([128] and references therein). The solvent free energy relationship can be further split into interactions with the polar head groups and interactions with the hydrophobic lipid chains. In this model, when "good" solvents are added, the interfacial area swells to increase the total contract with the solvent. For organic solvents that are water-miscible and have a high solubility for alkanes, such as acetone and ethanol, the interaction increases the interfacial area by reducing the interaction free energy between the solvent and the interfacial alkyl chains [128]. On the other hand, the interaction of "poor" solvents with lipids is unfavorable and has larger free energy penalty. As a result, the interfacial segments shrink in size to prevent contact with the solvent. Consequently, "good" solvents will favor interdigitation while "poor" solvents will destabilize the $L_\beta I$ phase.

The inhibition of interdigitation has also been described in terms of osmotic stress. Chemicals that apply osmotic stress, such as poly(ethylene glycol) tend to inhibit interdigitation [60]. As in the other models described above, this has been proposed to occur because of a decrease in the repulsive interaction between the lipid head groups.

Dimethyl sulfoxide (DMSO) is an example of a solvent inhibitor of interdigitation. The interaction of DMSO with membranes is of great interest because it can be used as a cryoprotectant for biological material, such as stem cells [129]. DMSO can also enhance the permeability of membranes [130]. The mechanism by which DMSO inhibits interdigitation is by decreasing the repulsion between head groups [131]. The ability of DMSO to form unusually strong hydrogen bonds may explain this effect [132]. This phenomenon can be clearly seen in the phase behavior of DHPC. Just as chemicals that favor interdigitation shift the pre-transition of DHPC to a higher temperature; factors that disfavor interdigitation shift the pre-transition to a lower temperature. The suppression of the pre-transition clearly demonstrates that DMSO destabilizes the $L_\beta I$ phase (Figure 13) [131].

A major caveat with these solvent models is that the interactions with lipids are often concentration-dependent. For instance, in the DPPC/DMSO/water system, three distinct effects are found within different DMSO concentration ranges [133]. Perhaps the most remarkable is at above mol fractions of ~0.9 DMSO, the T_m temperature is greatly elevated and an interdigitated gel phase is formed [133].

The disaccharide trehalose is another example of a chemical inhibitor of interdigitation [124]. Like DMSO, the interactions of trehalose with membranes show promise in the cryopreservation of biological material [129]. In the yeast *Saccharomyces cerevisiae*, trehalose appears to increase viability during ethanol fermentation and provide protection against oxidative stress [134-137]. Similar to DMSO, trehalose disfavors interdigitation by increasing the packing density of the lipid head groups [124,138,139]. However, there is disagreement over the exact molecular interaction with lipids. The main dispute is over whether or not sugars are directly bound to or excluded from the membrane surface [140,141]. Recently, Andersen et al. have tried to explain this discrepancy by proposing that there are two concentration-dependent interactions. In this explanation, trehalose binds strongly to the bilayer at low concentrations, but is gradually expelled above ~0.2 M [141].

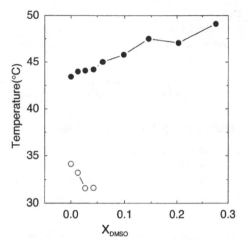

Figure 13. Phase transition temperatures of DHPC-MLV at various concentrations of DMSO (mole fraction) determined by DSC. (●) shows gel to liquid-crystalline phase transition temperatures and (○) shows $L_\beta I$ to $P_{\beta'}$ phase transition temperatures. Reprinted from Biochim. Biophys. Acta., 1467, Yamashita Y, Kinoshita K, Yamazaki M, Low concentration of DMSO stabilizes the bilayer gel phase rather than the interdigitated gel phase in dihexadecylphosphatidylcholine membrane, Pages No. 395-405, Copyright (2000), with permission from Elsevier [131].

2.13. The interdigitated gel phase versus the inverted hexagonal phase

A clear inverse relationship exists between the interdigitated phase gel phase and the inverted hexagonal phase (H_{II}) [56,128]. The major structural factor is the relative size of the lipid headgroup and the attraction/repulsion between headgroups. A lipid that forms the inverted hexagonal phase is unlikely to interdigitate and vice versa. The temperature dependence of these phases is also opposite. For example, with DHPC, the interdigitated phase is present only below the pre-transition. The interdigitated phase requires predominately *trans* confirmations in the hydrocarbon chains, so it is unlikely to form in the liquid crystalline phase where there are abundant *gauche* confirmations and a high degree of disorder [2,37]. In contrast, the inverted hexagonal phase typically forms well above the main transition into the liquid crystalline phase [8].

This relationship also extends to environmental factors that encourage or discourage interdigitation (Table 5). Chemicals that favor the interdigitated phase such as ethanol tend to destabilize the H_{II} phase [128,124 and references therein]. Interdigitation is favored because the surface area per lipid head group in the $L_\beta I$ phase is substantially larger versus non-interdigitated membranes [124]. The H_{II} is the opposite because it requires a small head group area. Solvents that stabilize the H_{II} phase like dimethyl sulfoxide therefore also inhibit interdigitation [56,122,128]. This relationship appears to apply to hydrostatic pressure as well. While increased pressure favors interdigitation (Tables 2 and 3), pressure destabilizes the inverted hexagonal phase in PE lipids [56].

$L_\beta I$	H_{II}
Large head group repulsion	Small head group repulsion
Chaotropic chemicals	Kosmotropic chemicals
High hydrostatic pressure	Low hydrostatic pressure
Low Temperatures	High Temperatures

Table 5. Factors that stabilize the interdigitated gel ($L_\beta I$) phase and the inverted hexagonal (H_{II}) phase.

2.14. Influence of hydration and pH on the $L_\beta I$ phase

While most interdigitated systems are studied in excess water, interdigitation can be affected at less than full hydration. For instance, interdigitation of DHPC is reliant on hydration, as coexisting interdigitated and non-interdigitated phases are found at low hydration [142,143]. However, the cationic EDPPC may be interdigitated in the dry state [97].

Furthermore, substituting deuterium oxide (D_2O) for water slightly disfavors the spontaneous interdigitated phase of DHPC [144]. Using D_2O also increases the threshold concentration for the chemically-induced interdigitation of DPPC [27] and increases threshold pressure for interdigitation [145]. These phenomena are explained by the different hydrophobic interactions and interfacial energies in H_2O versus D_2O [27,144,145].

Changing the pH of the aqueous solution can also affect interdigitation. In DHPC membranes a low pH will inhibit interdigitation [59]. As the pH is lowered the phosphate groups are protonated and ultimately the total repulsive force between head groups is decreased, disfavoring interdigitation [59]. The pH is also highly relevant to the interdigitated phase in charged lipids, such as PGs. At a high pH, the electrostatic repulsion between head groups that encourages interdigitation in PGs is increased [78].

2.15. Lipid mixtures and interdigitated/non-interdigitated gel phase coexistence

Under certain circumstances, interdigitated and non-interdigitated phases can coexist within a membrane even though the boundaries between these domains are considered to be energetically unfavorable [115,116]. The uneven structure between these domains can significantly increase the membrane permeability [146,147]. With the variety of lipids now known to interdigitate, there are many possible lipid systems that will have complex phase diagrams involving the $L_\beta I$ phase.

Gel phase coexistence can often be found in binary mixtures of a lipid that can spontaneously interdigitate (e.g. F-DPPC or EDPPC) and one that cannot (e.g. DPPC or PE lipids). For example, at equimolar amounts of F-DPPC and DPPC, interdigitated F-DPPC-rich domains create a phase-segregated system [69,117]. On DSC scans this manifests itself as multiple peaks (Figure 8). The peaks with the greatest transition hysteresis likely correspond to interdigitated domains rich in F-DPPC. When the F-DPPC molar fraction is large, the hysteresis is also increased [69]. Additionally, gel phase coexistence occurs in the mixture of 1,2-dielaidoyl-sn-glycero-3-phosphoethanolamine (DEPE) and EDPPC [95].

However, this is not true of all such binary mixtures. Outside of the phase transition regions in DPPC/EDPPC, for instance, there is no gel phase segregation [95].

Mixing an interdigitated lipid with cholesterol can also produce gel phase coexistence. Cholesterol-poor interdigitated domains and cholesterol-rich non-interdigitated domains have been found in DHPC/cholesterol [116], F-DPPC/cholesterol [117], and EDPPC/cholesterol [95]. For these mixtures, the lipids with the most stable interdigitated phase tend to have a larger region of phase coexistence within the phase diagram.

Alternatively, a lipid such as DPPC that can be chemically induced to interdigitate can be mixed with lipids that cannot, such as PE lipids [147]. The DPPC-rich domains will interdigitate with ethanol, but domains composed of mostly PE lipid will not. A similar result can be achieved in mixtures of DPPC/cholesterol/ethanol, where the cholesterol-rich domains remain non-interdigitated in the presence of ethanol [39,146].

It is also possible to have coexistence in membranes with only one lipid. For instance, coexisting interdigitated and non-interdigitated phases form in supported F-DPPC membranes where the lateral expansion of the lipid film is restricted [68]. This results in a "frustrated" state, where the energetically favorable interdigitated phase cannot fully form due to constraints in topology and the available surface area [68]. Additionally, while the 16-carbon chain length DPPG does not spontaneously interdigitate, the 18-carbon chain DSPG spontaneously forms an interdigitated gel phase that coexists with a non-interdigitated gel phase [78]. This two-phase coexistence was attributed to a kinetically trapped system that is not at thermal equilibrium [78].

2.16. Applications of the interdigitated gel phase

One of the most promising applications for the interdigitated gel phase is the creation of large unilamellar vesicles termed interdigitation-fusion (IF) vesicles [44,148]. Figure 14 demonstrates the process for the creation of IF liposomes using ethanol [148]. Below the main transition, the ethanol causes the formation rigid and flat interdigitated sheets [149]. These sheets are surprisingly stable under the T_m, even when ethanol is removed [149]. When the temperature is raised above the main transition, the sheets fuse into large vesicles. This fusion encapsulates particles from the surrounding solution [149,150]. These materials include other small vesicles, biological macromolecules, colloids, and nanoparticles [149,150]. The amphiphilic nature of lipids allows for the capture of hydrophobic materials [151]. Transmembrane insertion of protein into IF vesicles has also been achieved using electropulsation [152].

The IF procedure can also be used to create multicompartment vesicle-in-vesicle structures called "vesosomes" [150]. These multicompartment vesicles should be closer replicas of eukaryotic cells than regular vesicles [150,153]. Therefore, vesosomes have the potential to more closely mimic biological conditions and reactions in artificial cells [150,154,155]. Furthermore, the retention of encapsulated material can be substantially increased in vesosomes [151,156,157]. These vesicles are highly customizable because the composition of the inner and outer components can be varied [149,150,154]. As a result, it is theoretically possible to use vesosomes as controlled nanoreactors [153,155]. For complex and expensive chemistry such as enzyme reactions, vesosomes should be able to optimize reaction conditions and drastically reduce the amount of reagents needed [155].

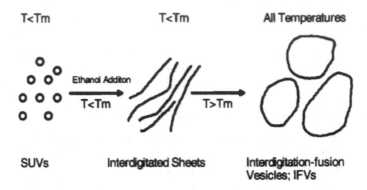

Figure 14. Liposome formation by interdigitation fusion (IF) using ethanol. Reprinted from Biochim. Biophys. Acta., 1195, Ahl PL, Chen L, Perkins WR, Minchey SR, Boni LT, Taraschi TF, Janoff AS, Interdigitation-fusion: a new method for producing lipid vesicles of high internal volume, Pages No. 237-244, Copyright (1994), with permission from Elsevier [148].

As described by Ahl et al. [44], there are four general guidelines for IF liposomes: (1) the lipids must be able to form the interdigitated phase; (2) the precursor liposomes should be small, preferably sonicated SUVs; (3) the temperature of the precursor SUV suspension after the addition of the alcohol must be below the T_m of the phospholipids; and (4) the temperature should be raised above the T_m of the phospholipids after the formation of the interdigitated sheets. Therefore, the creation of these liposomes is dependent on the lipid composition. Adding cholesterol and lipids containing *cis* double bonds can compromise the formation of IF liposomes [103,148]. PE lipids are also unsuitable because of their reluctance to interdigitate [44].

A similar result can be achieved using pressure to create pressure-induced fusion (PIF) liposomes [103]. An advantage of this technique is that no organic solvent is required and it is an effective sterilization method [103]. The captured volume of the IF or PIF vesicles is larger than other techniques for liposome preparation ([44] and references therein).

3. Conclusions

As an analytical instrument, DSC offers many advantages. One advantage is the simplicity of the sample preparation procedure. Samples do not have to be supported or spatially oriented and do not require the insertion of a membrane probe. For sensitive low enthalpy phase transitions, it is a great benefit not to need a probe so that the purity of the sample can be maintained. The importance of this can be seen in alcohol-induced interdigitation, where the low enthalpy pre-transition is an important aspect of the analysis (Figure 5) [15]. Moreover, the effects of pressure can be measured concomitantly with calorimetry data with the appropriate equipment. This greatly expands the range of the phase diagram that can be experimented with.

We have shown that DSC can accurately measure changes in the thermodynamic properties of phospholipid membranes with the addition of chemicals that either encourage or discourage interdigitation. DSC is particularly well-suited for the study of chemically-induced interdigitation because it is sensitive enough to detect small, incremental changes in phase transition temperatures (Figure 5). With the capability to perform heating and cooling scans at a constant rate, the transition hysteresis can also be easily determined. In addition, the transition enthalpy can highlight the "biphasic" behavior above and below the threshold concentration for interdigitation (Figure 6).

Moreover, DSC can reveal how changes in either the hydrocarbon chains (Figure 1) or in the polar head group (Figure 2) will affect the thermodynamics. Modifications that either encourage or discourage interdigitation are summarized in Figure 15. Understanding the importance of structural differences reveals the importance of lipid diversity in biological membranes. Lipid composition can help explain why, for example, a peptide might interact differently with human versus microbial membranes [81]. With the increasing popularity of liposomes for pharmaceutical applications and research, it also is essential to find suitable lipid candidates. For instance, calorimetry can be applied to screen potential IF vesicles by determining whether interdigitation is present and by determining the T_m temperature.

In addition, more information can be inferred from DSC data than the phase transition temperature. With careful analysis, the nature of the lipid/solvent interaction and the properties of the chemicals themselves can be derived. For example, the characteristics of kosmotropic and chaotropic chemicals are clearly reflected in their effects on lipid membranes (see section 3.12.). This analysis can also increase the understanding of how chemicals interact with biological membranes, such as why chemicals like DMSO and trehalose can protect cells during cryopreservation [129].

However, DSC also has limitations when analyzing phospholipid samples. Perhaps the greatest weakness is the lack of direct structural information. As a consequence, relying solely on DSC data can be misleading. For instance, the pre-transition peaks of DPPC and DHPC look similar on DSC thermograms. However, the actual nature of the transition is substantially different (Figure 3). While the structure can often be reasonably inferred from thermodynamic properties, it is not as robust as other experimental techniques [6]. Additionally, while alterations in the macromolecular structure can be reflected in DSC data (see section 3.10.), the changes are not specific enough to be able to infer the true structure.

Overlapping or multiple transitions can also present a problem. In F-DPPC/DPPC, the multiple peaks reflect the presence of phase segregation (Figure 8), but this is not always the case. Multiple DSC peaks can also indicate separate phase transitions that involve the entire membrane. In the case of EDPPC, different morphologies result in separate DSC peaks (Figure 11) [97]. Overlapping peaks can also obscure individual transitions, especially when there are multiple components in the membrane and the transition peaks are broad.

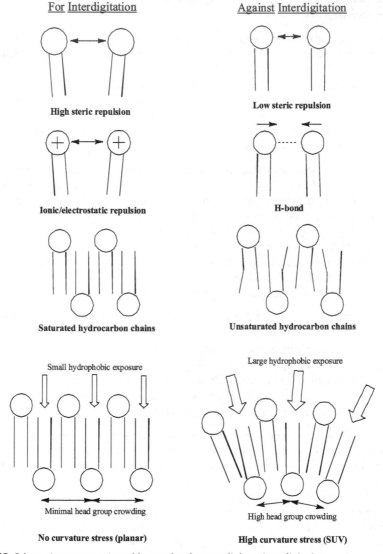

Figure 15. Schematic representation of factors that favor or disfavor interdigitation.

Fortunately, one of the greatest strengths of DSC data is that it is highly compatible with other analytical techniques. In the case of the $L_\beta I$ phase, methods such as x-ray diffraction, nuclear magnetic resonance, and fluorescence techniques can fill in the gaps ([6] and references therein). Additionally, DSC is highly valuable in determining the relevant temperature range to use for the other experimental techniques.

The stability of the interdigitated phase plainly demonstrates the balance of forces within the membrane. Factors as varied as electrostatic and steric interactions, van der Waals forces, solvent binding at the interface, and the presence of double bonds all contribute to the properties of hydrated phospholipid membranes. DSC provides a way to judge the resulting balance of these forces by measuring the stability of different thermodynamic phases. Consequently, the wealth of information calorimetric analysis provides ensures that DSC will remain an invaluable tool for the study of membrane biophysics.

Author details

Eric A. Smith and Phoebe K. Dea[*]
Department of Chemistry, Occidental College, Los Angeles, USA

Abbreviations

differential scanning calorimetry (DSC)
main transition temperature (T_m)
pre-transition temperature (T_P)
small unilamellar vesicle (SUV)
large unilamellar vesicle (LUV)
multilamellar vesicle (MLV)
interdigitation-fusion vesicle (IFV)
interdigitated gel phase ($L_\beta I$)
planar gel phase (L_β')
ripple gel phase (P_β')
liquid crystalline phase (L_α)
inverted hexagonal phase (H_{II})
crystalline bilayer phase (L_c)
liquid-ordered (l_o)
phosphatidylcholine (PC)
phosphatidylglycerol (PG)
phosphatidylethanolamine (PE)
phosphatidylethanol (Peth)
1,2-dipalmitoyl-*sn*-glycero-3-phosphocholine (DPPC)
1,3-dipalmitoyl-*sn*-glycero-2-phosphocholine (1,3-DPPC or β-DPPC)
1,2-di-*O*-hexadecyl-*sn*-glycero-3-phosphocholine (DHPC)
1-palmitoyl-2-(16-fluoropalmitoyl)*sn*-glycero-3-phosphocholine (F-DPPC)
1,2-dipalmitoyl-*sn*-glycero-3-phosphoethanol (DPPeth)

[*] Corresponding Author

1,2-dipalmitoyl-*sn*-glycero-3-phospho-(1'-*rac*-glycerol) (DPPG)
1-palmitoyl-2-hydroxy-*sn*-glycero-3-phosphocholine (16:0 LPC)
1,2-dipalmitoyl-*sn*-glycero-3-ethylphosphocholine (EDPPC or Et-DPPC)
1,2-dipalmitoyl-*sn*-glycero-3-phosphoethanolamine (DPPE)
1,2-di-*O*-hexadecyl-*sn*-glycero-3-phosphoethanolamine (DHPE)
1,2-dielaidoyl-*sn*-glycero-3-phosphoethanolamine (DEPE)
1,2-dioleoyl-*sn*-glycero-3-phosphocholine (DOPC)
1,2-dielaidoyl-*sn*-glycero-3-phosphocholine (DEPC)
1-stearoyl-2-oleoyl-phosphatidylcholine (SOPC)
1-palmitoyl-2-oleoyl-phosphatidylcholine (POPC)
1-palmitoyl-2-oleoyl-phosphatidylglycerol (POPG)

4. References

[1] Mabrey S, Sturtevant JM (1976) Investigation of Phase Transitions of Lipids and Lipid Mixtures by High Sensitivity Differential Scanning Calorimetry. Proc. natl. acad. sci. USA. 73: 3862-3866.

[2] McElhaney RN (1982) The Use of Differential Scanning Calorimetry and Differential Thermal Analysis in Studies of Model and Biological Membranes. Chem. phys. lipids. 30: 229-259.

[3] Chiu MH, Prenner EJ (2011) Differential Scanning Calorimetry: An Invaluable Tool for a Detailed Thermodynamic Characterization of Macromolecules and their Interactions. J. pharm. bioallied sci. 3: 39-59.

[4] Demetzos C (2008) Differential Scanning Calorimetry (DSC): A Tool to Study the Thermal Behavior of Lipid Bilayers and Liposomal Stability. J. liposomal res. 18: 159-173.

[5] Jackson MB, Sturtevant JM (1977) Studies of the Lipid Phase Transitions of *Escherichia coli* by High Sensitivity Differential Scanning Calorimetry. J. biol. chem. 252: 4749-4751.

[6] Slater JL, Huang CH (1988) Interdigitated Bilayer Membranes. Prog. lipid res. 27: 325–359.

[7] Koynova R, Caffrey M (1998) Phases and Phase Transitions of the Phosphatidylcholines. Biochim. biophys. acta. 1376: 91-145.

[8] Koynova R, Caffrey M (1994) Phases and Phase-Transitions of the Hydrated Phosphatidylethanolamines. Chem. phys. lipids 69: 1-34.

[9] Koynova R, Caffrey M (2002) An Index of Lipid Phase Diagrams. Chem. phys. lipids. 115: 107-219.

[10] Marsh D (2010) Structural and Thermodynamic Determinants of Chain-Melting Transition Temperatures for Phospholipid and Glycolipids Membranes. Biochim. biophys. acta. 1789: 40-51.

[11] Xu H, Huang CH (1987) Scanning Calorimetric Study of Fully Hydrated Asymmetric Phosphatidylcholines with One Acyl Chain Twice as Long as the Other. Biochem. 26: 1036-1043.

[12] Huang C, McIntosh TJ (1997) Probing the Ethanol-Induced Chain Interdigitation in Gel-State Bilayers of Mixed-Chain Phosphatidylcholines. Biophys. j. 70: 2702-2709.

[13] Löbbecke L, Cevc G (1995) Effects of Short-Chain Alcohols on the Phase Behavior and Interdigitation of Phosphatidylcholine Bilayer Membranes. Biochim. biophys. acta. 1237: 59-69.

[14] Wang Y, Dea P (2009) Interaction of 1-propanol and 2-propanol with Dipalmitoylphosphatidylcholine Bilayer: A Calorimetric Study. J. chem. eng. data. 54: 1447-1451.

[15] Reeves MD, Schawel AK, Wang W, Dea P (2007) Effect of Butanol Isomers on Dipalmitoylphosphatidylcholine Bilayer Membranes. Biophys. chem. 128: 13-18.

[16] Griffin KL, Cheng C-Y, Smith EA, Dea PK (2010) Effects of Pentanol Isomers on the Phase Behavior of Phospholipid Bilayer Membranes. Biophys. chem. 152: 178-183.

[17] Hata T, Matsuki H, Kaneshina S (2000) Effect of Local Anesthetics on the Phase Transition Temperatures of Ether- and Ester-Linked Phospholipid Bilayer Membranes. Colloid surf. b. biointer. 18: 41-50.

[18] Maruyama S, Hata T, Matsuki H, Kaneshina S (1997) Effects of Pressure and the Local Anesthetic Tetracaine on Dihexadecylphosphatidylcholine Bilayer Membrane. Coll. surf. b. 8: 261-266.

[19] McIntosh TJ, McDaniel RV, Simon SA (1983) Induction of an Interdigitated Gel Phase in Fully Hydrated Phosphatidylcholine Bilayers. Biochim. biophys. acta. 731: 109-114.

[20] Matsingou C, Demetzos C (2007) Calorimetric Study on the Induction of Interdigitated Phase in Hydrated DPPC Bilayers by Bioactive Labdanes and Correlation to their Liposomal Stability: The Role of Chemical Structure. Chem. phys. lipids. 145: 45-62.

[21] Potamitis C, Chatzigeorgiou P, Siapi E, Viras K, Mavromoustakos T, Hodzic A, Pabst G, Cacho-Nerin F, Laggner P, Rappolt M (2011) Interactions of the AT1 Antagonist Valsartan with Dipalmitoyl-phosohatidylcholine Bilayers. Biochim. biophys. acta. 1808: 1753-1763.

[22] Swamy MJ, Marsh D (1995) Thermodynamics of Interdigitated Phases of Phosphatidylcholine in Glycerol. Biophys. j. 69: 1402-1408.

[23] Boggs JM, Rangaraj G, Watts A (1989) Behavior of Spin Labels in a Variety of Interdigitated Lipid Bilayers. Biochim. biophys. acta biomembr. 981: 243-253.

[24] Boggs JM, Tümmler B (1993) Interdigitated Gel Phase Bilayers Formed by Unsaturated Synthetic and Bacterial Glycerolipids in the Presence of Polymyxin B and Glycerol. Biochim. biophys. acta. 1145: 42-50.

[25] Yamazaki M, Ohshika M, Kashiwagi N, Asano T (1992) Phase Transitions of Phospholipid Vesicles under Osmotic Stress and in the Presence of Ethylene Glycol. Biophys. chem. 43: 29-37.

[26] Kinoshita K., Asano T, Yamazaki M (1997) Interaction of the Surface of Biomembrane with Solvents: Structure of Multilamellar Vesicles of Dipalmitoylphosphatidylcholine in Acetone-Water Mixtures. Chem. phys. lipids. 85: 53-65.

[27] Kinoshita K, Yamazaki M. (1996) Organic Solvents Induce Interdigitated Gel Structures in Multilamellar Vesicles of Dipalmitoylphosphatidylcholine. Biochim. biophys. acta biomembr. 1284: 233-239.

[28] Wu F-G, Wang N-N, Tao L-F, Yu Z-W (2010) Acetonitrile Induces Nonsynchronous Interdigitation and Dehydration of Dipalmitoylphosphatidylcholine Bilayers. J. phys. chem. b. 114: 12685-12691.

[29] Cunningham BA, Tamura-Lis W, Lis LJ, Collins JM (1989) Thermodynamic Properties of Acyl Chain and Mesophase Transition for Phospholipids in KSCN. Biochim. biophs. acta biomembr. 984: 109-112.

[30] Cunningham BA, Quinn PJ, Wolfe DH, Tamura-Lis A, Lis LJ, Kucuk O, Westerman MP (1995) Real-Time X-ray Diffraction Study at Different Scan Rates of Phase Transitions for Dipalmitoylphosphatidylcholine in KSCN. Biochim. biophs. acta biomembr. 1233: 68-74.

[31] Tamai N, Matsui T, Moribayashi N, Goto M, Matsuki H, Kaneshina S (2008) Cholesterol Suppresses Pressure-Induced Interdigitation of Dipalmitoylphosphatidylcholine Bilayer Membrane. Chem. lett. 37: 604-605.

[32] Zeng J, Chong PLG (1991) Interactions between Pressure and Ethanol on the Formation of Interdigitated DPPC Liposomes: A Study with Prodan Fluorescence. Biochem. 30: 9485-9491.

[33] Braganza LF, Worcester DL (1986) Hydrostatic Pressure Induces Hydrocarbon Chain Interdigitation in Single-Component Phospholipid Bilayers. Biochem. 25: 2591-2596.

[34] Tamai N, Goto M, Matsuki H, Kaneshina S (2010) A Mechanism of Pressure-Induced Interdigitation of Lipid Bilayers. J. phys. conf. ser. 215: 012161 1-7.

[35] Hui SW, Huang C-H (1986) X-ray Diffraction Evidence for Fully Interdigitated Bilayers of 1-stearoyllysophosphatidylcholine. Biochem. 25: 1330-1335.

[36] Biltonen RL, Lichtenberg D (1993) The Use of Differential Scanning Calorimetry as a Tool to Characterize Liposome Preparations. Chem. phys. lipids. 64: 129-142.

[37] Heerklotz H (2004) The Microcalorimetry of Lipid Membranes. J. phys. condens. matter 16: R441-R467.

[38] Zeng J, Smith K, Chong PL (1993) Effects of Alcohol-Induced Lipid Interdigitation on Proton Permeability in L-α-Dipalmitoylphosphatidylcholine Vesicles. Biophys. j. 65: 1404-1414.

[39] Tierney KJ, Block DE, Longo ML (2005) Elasticity and Phase Behavior of DPPC Membrane Modulated by Cholesterol, Ergosterol, and Ethanol. Biophys. j. 89: 2481-2493.

[40] Barry JA, Gawrisch K (1995) Effects of Ethanol on Lipid Bilayers Containing Cholesterol, Gangliosides, and Sphingomyelin. Biochem. 34: 8852-8860.

[41] Kõiv A, Kinnunen PKJ (1992) Influence of Ca²⁺ and Ethanol on the Aggregation and Thermal Phase Behavior of 1-dihecadecylphosphatidylcholine Liposomes. Chem. phys. lipids. 62: 253-261.

[42] Vierl U, Löbbecke L, Nagel N, Cevc G (1994) Solute Effects on the Colloidal and Phase Behavior of Lipid Bilayer Membranes: Ethanol-dipalmitoylphosphatidylcholine Mixtures. Biophys. j. 67: 1067-1079.

[43] Nagel NE, Cevc G, Kirchner S (1992) The Mechanism of the Solute-Induced Chain Interdigitation in Phosphatidylcholine Vesicles and Characterization of the Isothermal

Phase Transitions by Means of Dynamic Light Scattering. Biochim. biophys. acta biomembr. 1111: 263-269.

[44] Ahl PL, Perkins WR (2003) Interdigitation-Fusion Liposomes. Methods enzymol. 367: 80-98.

[45] Adachi T, Takahashi H., Ohki K, Hatta I (1995) Interdigitated Structure of Phospholipid-Alcohol Systems Studied by X-ray Diffraction. Biophys. j. 68: 1850-1855.

[46] Simon SA, McIntosh TJ (1984) Interdigitated Hydrocarbon Chain Packing Causes the Biphasic Transition Behavior in Lipid/alcohol Suspensions. Biochim. biophys. acta. 773: 169-172.

[47] Rowe ES (1983) Lipid Chain Length and Temperature Dependence of Ethanol-Phosphatidylcholine Interaction. Biochem. 22: 3299-3305.

[48] Rowe ES (1985) Thermodynamic Reversibility of Phase Transitions: Specific Effects of Alcohols on Phosphatidylcholines. Biochim. biophys. acta. 813: 321-330.

[49] Veiro JA, Nambi P, Rowe ES (1988) Effect of Alcohols on the Phase Transitions of Dihexadecylphosphatidylcholine. Biochim. biophys. acta biomembr. 943: 108-111.

[50] Singh H, Emberley J, Morrow MR (2008) Pressure Induces Interdigitation Differently in DPPC and DPPG. Eur. biophys. j. 37: 783-792.

[51] Matsuki H, Miyazaki E, Sakano F, Tamai N, Kaneshina S (2007) Thermotropic and Barotropic Phase Transitions in Bilayer Membranes of Ether-linked Phospholipids with Varying Alkyl Chain Lengths. Biochim. biophys. acta. 1768: 479-489.

[52] Laggner P, Lohner K, Degovics G, Müller, Schuster KA (1987) Structure and Thermodynamics of the Dihexadecylphosphatidylcholine–Water System. Chem. phys. lipids. 44: 31-60.

[53] Kim JT, Mattai J, Shipley GG (1987) Bilayer Interactions of Ether- and Ester-Linked Phospholipids: Dihexadecyl- and Dipalmitoylphosphatidylcholines. Biochem. 26: 6599-6603.

[54] Ichimori H, Hata T, Matsuki H, Kaneshina S (1998) Barotropic Phase Transitions and Pressure-induced Interdigitation on Bilayer Nembranes of Phospholipids with Varying Acyl Chain Lengths, Biochim. biophys. acta biomembr. 1414: 165-174.

[55] Kusube M, Matsuki H, Kaneshina S (2005) Thermotropic and Barotropic Phase Transitions of N-methylated Dipalmitoylphosphatidylethanolamine Bilayers. Biochim. biophys. acta biomembr. 1668: 25-32.

[56] Cheng A, Mencke A, Caffrey M (1996) Manipulating Mesophase Behavior of Hydrated DHPE: An X-ray Diffraction Study of Temperature and Pressure Effects. J. phys. chem. 100: 299-306.

[57] Lewis RNAH, Pohle W, McElhaney RN (1996) The Interfacial Structure of Phospholipid Bilayers: Differential Scanning Calorimetry and Fourier Transform Infrared Spectroscopic Studies of 1,2-dipalmitoyl-sn-glycero-3-phosphorylcholine and its Dialkyl and Acyl-alkyl Analogs. Biophys. j. 70: 2736-2746.

[58] Haas NS, Sripada PK, Shipley GG (1990) Effect of Chain-linkage on the Structure of Phosphatidylcholine Bilayers. Biophys. j. 57: 117-124.

[59] Furuike S, Levadny VG, Li SJ, Yamazaki M (1999) Low pH Induces an Interdigitated Gel to Bilayer Gel Phase Transition in Dihexadecylphosphatidylcholine Membrane, Biophys. j. 77: 2015-2023.

[60] Hatanaka Y, Kinoshita K, Yamazaki M (1997) Osmotic Stress Induces a Phase Transition from Interdigitated Gel Phase to Bilayer Gel Phase in Multilamellar Vesicles of Dihexadecylphosphatidylcholine, Biophys. chem. 65: 229-233.

[61] Kaneshina S, Maruyama S, Matsuki H (1996) Effect of Pressure on the Phase Behavior of Ester- and Ether-linked Phospholipid Bilayer Membranes. Prog. biotechnol. 13: 175-180.

[62] Hing FS, Maulik PR, Shipley GG (1991) Structure and Interactions of Ether- and Ester-linked Phosphatidylethanolamines. Biochem. 30: 9007-9015.

[63] Serrallach EN, Dijkman R, de Haas GH, Shipley GG (1983) Structure and Thermotropic Properties of 1,3-dipalmitoyl-glycero-2-phosphocholine. J. mol. biol. 170: 155-174.

[64] Dluhy RA, Chowdhry BZ, Cameron DG (1985) Infrared Characterization of Conformational Differences in the Lamellar Phases of 1,3-dipalmitoyl-sn-glycero-2-phosphocholine. Biochim. biophys. acta biomembr. 821: 437-444.

[65] Seelig J, Dijkman R, de Haas GH (1980) Thermodynmaic and Conformational Studies on sn-2-phosphatidylcholines in Monolayers and Bilayers. Biochem. 19: 2215-2219.

[66] Cunningham BA, Midmore L, Kucuk O, Lis LJ, Westerman MP, Bras W, Wolfe DH, Quinn PJ, Qadri SB (1995) Sterols Stabilize the Ripple Phase Structure in Dihexadecylphosphatidylcholine. Biochim. biophys. acta. 1233: 75-83.

[67] Hirsh DJ, Lazaro N, Wright LR, Boggs JM, McIntosh TJ, Schaefer J, Blazyk J (1998) A New Monofluorinated Phosphatidylcholine Forms Interdigitated Bilayers. Biophys. j. 75: 1858-1868.

[68] Sanii B, Szmodis AW, Bricarello DA, Oliver AE, Parikh AN (2010) Frustrated Phase Transformations in Supported, Interdigitating Lipid Bilayers. J. phys. chem. b. 114: 215-219.

[69] Smith EA, van Gorkum CM, Dea PK (2010) Properties of Phosphatidylcholine in the Presence of its Monofluorinated Analogue. Biophys. chem. 147: 20-27.

[70] McDonough B, Macdonald PM, Sykes BD, McElhaney RN (1983) Fluorine-19 Nuclear Magnetic Resonance Studies of Lipid Fatty Acyl Chain Order and Dynamics in Acholeplasma laidlawii B Membranes. A Physical, Biochemical, and Biological Evaluation of Monofluoropalmitic Acids as Membrane Probes. Biochem. 22: 5097-5103.

[71] Santaella C, Vierling P, Riess JG, Gulik-Krzywicki T, Gulik A, Monasse B (1994) Polymorphic Phase Behavior of Perfluoroalkylated Phosphatidylcholines. Biochim. biophys. acta. 1190: 25-39.

[72] McIntosh TJ, Simon SA, Vierling P, Santaella C, Ravily V (1996) Structure and Interactive Properties of Highly Fluorinated Phospholipid Bilayers. Biophys. j. 71: 1853-1868.

[73] Barbarich TJ, Rithner CD, Miller SM, Anderson OP, Strauss SH (1999) Significant Inter- and Intramolecular O−H···FC Hydrogen Bonding. J. am. chem. soc. 121: 4280-4281.

[74] Caminati W, Melandri S, Maris A, Ottaviani P (2006) Relative Strengths of the O−H···Cl and O−H···F Hydrogen Bonds. Angew. chem. int. ed. 45: 2438-2442.

[75] Hyla-Kryspin I, Haufe G, Grimme S (2004) Weak Hydrogen Bridges: A Systematic Theoretical Study on the Nature and Strength of C—H···F—C Interactions. Chem. eur. j. 10: 3411-3422.

[76] O'Hagan D (2008) Understanding Organofluorine Chemistry: An Introduction to the C—F bond. Chem. soc. rev. 37: 308-319.

[77] Toimil P, Prieto G, Miñones Jr. J, Sarmiento F (2010) A Comparative Study of F-DPPC/DPPC Mixed Monolayers: Influence of Subphase Temperature on F-DPPC and DPPC Monolayers. Phys. chem. chem. phys. 12: 13323-13332.

[78] Pabst G, Danner S, Karmakar S, Deutsch G, Raghunathan VA (2007) On the Propensity of Phosphatidylglycerols to Form Interdigitated Phases. Biophys. j. 93: 513-525.

[79] Wang P-Y, Lu J-Z, Chen J-W, Hwang F (1994) Interaction of the Interdigitated DPPG or DPPG/DMPC Bilayer with Human Erythrocyte Band 3: Differential Scanning Calorimetry and Fluorescence Studies. Chem. phys. lipids. 69: 241-249.

[80] Boggs JM, Rangaraj G (1997) Greater Partitioning of Small Spin Labels into Interdigitated than into Non-interdigitated Gel Phase Bilayers. Chem. phys. lipids. 87: 1-15.

[81] Sevcsik E, Pabst G, Jilek A, Lohner K (2007) How Lipids Influence the Mode of Action of Membrane-active Peptides. Biochim. biophys. acta. 1768: 2586-2595.

[82] Sevcsik E, Pabst G, Richter W, Danner S, Amenitsch H, Lohner K (2008) Interaction of LL-37 with Model Membrane Systems of Different Complexity: Influence of the Lipid Matrix. Biophys. j. 94: 4688-4699.

[83] Pabst G, Grage SL, Danner-Pongratz S, Jing W, Ulrich AS, Watts A, Lohner K, Hickel A (2008) Membrane Thickening by the Antimicrobial Peptide PGLa. Biophys. j. 95: 5779-5788.

[84] Bondar OP, Rowe ES (1996) Thermotropic Properties of Phosphatidylethanols. Biophys. j. 71: 1440-1449.

[85] Wilkinson DA, Tirrell DA, Turek AB, McIntosh TJ (1987) Tris Buffer Causes Acyl Chain Interdigitation in Phosphatidylglycerol. Biochim. biophys. acta. 905: 447-453.

[86] Ranck JL, Tocanne JF (1982) Choline and Acetylcholine Induce Interdigitation of Hydrocarbon Chains in Dipalmitoylphosphatidylglycerol Lamellar Phase with Stiff Chains. FEBS lett. 143: 171-174.

[87] Hao Y-H, Xu Y-M, Chen J-W, Huang F (1998) A Drug-Interaction Model: Atropine Induces Interdigitated Bilayer Structure. Biochem. biophys. res. commun. 245: 439-442.

[88] Boon JM, McClain RL, Breen JJ, Smith BD (2001) Inhibited Phospholipid Translocation Across Interdigitated Phosphatidylglycerol Vesicle Membranes. J. supramol. chem. 1: 17-21.

[89] Wang P-Y, Chen J-W, Hwang F (1993) Anisodamine Causes Acyl Chain Interdigitation in Phosphatidylglycerol. FEBS lett. 332: 193-196.

[90] Fang J, Barcelona MJ, Alvarez PJJ (2000) A Direct Comparison Between Fatty Acid Analysis and Intact Phospholipid Profiling for Microbial Identification. Org. geochem. 31: 881-887.

[91] Lohner K, Blondelle SE (2005) Molecular Mechanisms of Membrane Perturbation by Antimicrobial Peptides and the Use of Biophysical Studies in the Design of Novel Peptide Antibiotics. Comb. chem. high throughput screen. 8: 241-246.

[92] Gustavsson L, Alling C (1987) Formation of Phosphatidylethanol in Rat Brain by Phospholipase D. Biochem. biophys. res. commun. 142: 958-963.

[93] Gustavsson E (1995) Phosphatidylethanol Formation: Specific Effects of Ethanol Mediated via Phospholipase D. Alcohol alcoholism. 30: 391-406.

[94] Danner S, Pabst G, Lohner K, Hickel A (2008) Structure and Thermodynamic Behavior of *Staphylococcus aureus* Lipid Lysyl-dipalmitoylphosphatidylglycerol, Biophys. j. 94: 2150-2159.

[95] Koynova R, MacDonald RC (2003) Mixtures of Cationic Lipid *O*-ethylphosphatidylcholine with Membrane Lipids and DNA: Phase Diagrams, Biophys. j. 85: 2449-2465.

[96] Lewis RNAH, Winter I, Kriechbaum M, Lohner K, McElhaney RN (2001) Studies of the Structure and Organization of Cationic Lipid Bilayer Membranes: Calorimetric, Spectroscopic, and X-ray Diffraction Studies of Linear Saturated P-O-ethyl Phosphatidylcholines, Biophys. j. 80: 1329-1342.

[97] Koynova R, RC MacDonald (2003) Cationic O-ethylphosphatidylcholines and their Lipoplexes: Phase Behavior Aspects, Structural Organization and Morphology. Biochim. biophys. acta. 1613: 39-48.

[98] MacDonald RC, Ashley GW, Shida MM, Rakhmanova VA, Tarahovshy YS, Pantazatos DP, Kennedy MT, Pozharski EV, Baker KA, Jones RD, Rosenzweig HS, Choi KL, Qiu R, McIntosh TJ (1999) Physical and Biological Properties of Cationic Triesters of Phosphatidylcholine. Biophys. j. 77: 2612-2629.

[99] Koynova R, MacDonald RC (2004) Columnar DNA Superlattices in Lamellar *O*-Ethylphosphatidylcholine Lipoplexes: Mechanism of the Gel-liquid Crystalline Lipid Phase Transition. Nano lett. 4: 1475-1479.

[100] Kennedy MT, Pozharski EV, Rakmanova VA, MacDonald RC (2000) Factors Governing the Assembly of Cationic Phospholipid-DNA Complexes. Biophys. j. 78: 1620-1633.

[101] MacDonald RC, Rakhmanova VA, Choi KL, Rosenzweig HS, Lahiri MK (1999) *O*-ethylphosphatidylcholine: A Metabolizable Cationic Phospholipid which is a Serum-Compatible DNA Transfection Agent. J. pharm. sci. 88: 896-904.

[102] McIntosh TJ (1996) Hydration Properties of Lamellar and Non-lamellar Phases of Phosphatidylcholine and Phosphatidylethanolamine. Chem. phys. lipids. 81: 117-131.

[103] Perkins WR, Dause R, Li X, Davis TS, Ahl PL, Minchey SR, Taraschi TF, Erramilli S, Gruner SM, Janoff AS (1995) Pressure Induced Fusion (Pif) Liposomes: A Solventless Sterilizing Method for Producing Large Phospholipid Vesicles. J. liposome res. 5: 605-626.

[104] Tada K, Goto M, Tamai N, Matsuki H, Kaneshina S (2010) Pressure Effect on the Bilayer Phase Transition of Asymmetric Lipids with an Unsaturated Acyl Chain. Ann. N.Y. acad. sci. 1180: 77-85.

[105] Ichimori H, Hata T, Matsuki H, Kaneshina S (1999) Effect of Unsaturated Acyl Chains on the Thermotropic and Barotropic Phase Transitions of Phospholipid Bilayer Membranes. Chem. phys. lipids. 100: 151-164.

[106] Dalton LA, Miller KW (1993) *Trans*-unsaturated Lipid Dynamics: Modulation of Dielaidoylphosphatidylcholine Acyl Chain Motion by Ethanol. Biophys. j. 65: 1620-1631.

[107] Vanegas JM, Contreras MF, Faller R, Longo ML (2012) Role of Unsaturated Lipid and Ergosterol in Ethanol Tolerance of Model Yeast Biomembranes. Biophys. j. 102: 507-516.

[108] Mills TT, Huang J, Feigenson GW, Nagle JF (2009) Effects of Cholesterol and Unsaturated DOPC Lipid on Chain Packing of Saturated Gel-phase DPPC Bilayers. Gen. physiol. biophys. 28: 126-139.

[109] McIntosh TJ, Lin H, Li S, Huang C-H (2001) The Effect of Ethanol on the Phase Transition Temperature and the Phase Structure of Monounsaturated Phosphatidylcholines. Biochim. biophys. acta 1510: 219-230.

[110] Polozova A, Li X, Shangguan T, Meers P, Schuette DR, Ando N, Gruner SM, Perkins WR (2005) Formation of Homogeneous Unilamellar Liposomes from an Interdigitated Matrix. Biochim. biophys. acta. 1668: 117-125.

[111] Boni LT, Minchey SR, Perkins WR, Ahl PL, Slater JL, Tate MW, Gruner SM, Janoff AS (1993) Curvature Dependent Induction of the Interdigitated Gel Phase in DPPC Vesicles. Biochim. biophys. acta. 1146: 247-257.

[112] Komatsu H, Guy PT, Rowe ES (1993) Effect of Unilamellar Vesicle Size on Ethanol-Induced Interdigitation in Dipalmitoylphosphatidylcholine. Chem. phys. lipids. 65: 11-21.

[113] Mason JT, Huang CH, Biltonen RL (1983) Effect of Liposomal Size on the Calorimetric Behavior of Mixed-chain Phosphatidylcholine Bilayer Dispersions. Biochem. 22: 2013-2018.

[114] Komatsu H, Rowe ES (1991) Effect of Cholesterol on the Ethanol-Induced Interdigitated Gel Phase in Phosphatidylcholine: Use of Fluorophore Pyrene-Labeled Phosphatidylcholine. Biochem. 30: 2463-2470.

[115] Bondar OP, Rowe ES (1998) Role of Cholesterol in the Modulation of Interdigitation in Phosphatidylethanols. Biochim. biophys. acta. 1370: 207-217.

[116] Laggner P, Lohner K, Koynova R, Tenchov B (1991) The Influence of Low Amounts of Cholesterol on the Interdigitated Gel Phase of Hydrated Dihexadecylphosphatidylcholine. Chem. phys. lipids. 60: 153-161.

[117] Smith EA, Wang W, Dea PK (2012) Effects of Cholesterol on Phospholipid Membranes: Inhibition of the Interdigitated Gel Phase of F-DPPC and F-DPPC/DPPC. Chem. phys. lipids. 165: 151-159.

[118] Lu JZ, Hao YH, Chen JW (2001) Effect of Cholesterol on the Formation of an Interdigitated Gel Phase in Lysophosphatidylcholine and Phosphatidylcholine Binary Mixtures. J. biochem. 129: 891-898.

[119] McMullen TPW, Lewis RNAH, McElhaney RN (1994) Comparative Differential Scanning Calorimetric and FTIR and ^{31}P-NMR Spectroscopic Studies of the Effects of

Cholesterol and Androstenol on the Thermotropic Phase Behavior and Organization of Phosphatidylcholine Bilayers. Biophys. j. 66: 741-752.

[120] Clarke JA, Heron AH, Seddon JM, Law RV (2006) The Diversity of the Liquid Ordered (Lo) Phase of Phosphatidylcholine/Cholesterol Membranes: A Variable Temperature Multinuclear Solid-state NMR and X-ray Diffraction Study. Biophys. j. 90: 2383-2393.

[121] Rand RP, Pangborn WA, Purdon AD, Tinker DO (1975) Lysolecithin and Cholesterol Interact Stoichiometrically Forming Bimolecular Lamellar Structures in the Presence of Excess Water. Can. j. biochem. 53: 189-195.

[122] [122] Koynova R, Brankov J, Tenchov B (1997) Modulation of Lipid Phase Behavior by Kosmotropic and Chaotropic Solutes. Eur. biophys. j. 25: 261-274.

[123] Yu Z-W, Quinn PJ (1995) Phase Stability of Phosphatidylcholines in Dimethylsulfoxide Solutions. Biophys. j. 69: 1456-1463.

[124] Takahashi H, Ohmae H, Hatta I (1997) Trehalose-induced Destabilization of Interdigitated Gel Phase in Dihexadecylphosphatidylcholine. Biophys. j. 73: 3030-3038.

[125] Söderlund T, Alakoskela JM, Pakkanen AL, Kinnunen PKJ (2003) Comparison of the Effects of Surface Tension and Osmotic Pressure in the Interfacial Hydration of a Fluid Phospholipid Bilayer. Biophys. j. 85: 2333-2341.

[126] Luu DV, Cambon L, Mathlouthi M (1990) Perturbation of Liquid-Water Structure by Ionic Substances. J. mol. struct. 237: 411-419.

[127] Collins KD (1997) Charge Density-dependent Strength of Hydration and Biological Structure. Biophys. j. 72: 65-76.

[128] Kinoshita K, Li SJ, Yamazaki M (2001) The Mechanism of the Stabilization of the Hexagonal II (HII) Phase in Phosphatidylethanolamine Membranes in the Presence of Low Concentrations of Dimethyl Sulfoxide. Eur. biophys. j. 30: 207-220.

[129] Scheinkönig C, Kappicht S, Kolb H-J, Schleuning M (2004) Adoption of Long-term Cultures to Evaluate the Cryoprotective Potential of Trehalose for Freezing Hematopoietic Stem Cells. Bone marrow transplant. 34: 531-536.

[130] Notman R, Noro M, O'Malley B, Anwar J (2006) Molecular Basis for Dimethylsulfoxide (DSMO) Action on Lipid Membranes. J. am. chem. soc. 128: 13982-13983.

[131] Yamashita Y, Kinoshita K, Yamazaki M (2000) Low Concentration of DMSO Stabilizes the Bilayer Gel Phase Rather than the Interdigitated Gel Phase in Dihexadecylphosphatidylcholine Membrane. Biochim. biophys. acta. 1467: 395-405.

[132] Yu Z-W, Chen L, Sun S-Q, Noda I (2002) Determination of Selective Molecular Interactions Using Two-dimensional Correlation FT-IR Spectroscopy. J. phys. chem. a. 106: 6683-6687.

[133] Gordeliy VI, Kiselev MA, Lesieur P, Pole AV, Teixeira J (1998) Lipid Membrane Structure and Interactions in Dimethyl Sulfoxide/Water Mixtures. Biophys. j. 75: 2343-2351.

[134] Mansure JJ, Souza RC, Panek AD (1997) Trehalose Metabolism in *Saccharomyces cerevisiae* During Alcoholic Fermentation. Biotechnol. lett. 19: 1201-1203.

[135] Lucero P, Peñalver E, Moreno E, Lagunas R (2000) Internal Trehalose Protects Endocytosis from Inhibition by Ethanol in *saccharomyces cerevisiae*. Appl. environ. microbiol. 66: 4456-4461.

[136] Gibson BR, Lawrence SJ, Leclaire JPR, Powell CD, Smart KA (2007) Yeast Responses to Stresses Associated with Industrial Brewery Handling. FEMS microbiol. rev. 31: 535-569.

[137] Trevisol ETV, Panek AD, Mannarino SC, Eleutherio ECA (2011) The Effect of Trehalose on the Fermentation Performance of Aged Cells of *Saccharomyces cerevisiae*. Appl. microbiol. biotechnol. 90: 697-704.

[138] Nishiwaki T, Sakurai M, Inoue Y, Chŭjŏ R, Koybayashi S (1990) Increasing Packing Density of Hydrated Dipalmitoylphosphatidylcholine Unilamellar Vesicles Induced by Trehalose. Chem. lett. 19: 1841-1844.

[139] di Gregorio GM, Mariani P (2005) Rigidity and Spontaneous Curvature of Lipidic Monolayers in the Presence of Trehalose: A Measurement in the DOPE Inverted Hexagonal Phase. Eur Biophys. j. 34: 67-81.

[140] Villarreal MA, Díaz SB, Disalvo EA, Montich GG (2004) Molecular Dynamics Simulation Study of the Interaction of Trehalose with Lipid Membranes. Langmuir. 20: 7844-7851.

[141] Andersen HD, Wang C, Arleth L, Peters GH, Westh P (2011) Reconciliation of Opposing Views on Membrane-Sugar Interactions. Proc. natl. acad. sci. 108: 1874-1878.

[142] Kim JT, Mattai J, Shipley GG (1987) Gel Phase Polymorphism in Ether-linked Dihexadecylphosphatidylcholine Bilayers. Biochem. 26: 6592-6598.

[143] Laggner P, Lohner K, Degovics G, Müller K, Schuster A (1987) Structure and Thermodynamics of the Dihexadecylphosphatidylcholine-Water System. Chem. phys. lipids. 44: 31-60.

[144] Ohki K (1991) Effect of Substitution of Hydrogen Oxide by Deuterium Oxide on Thermotropic Transition Between the Interdigitated Gel phase and the Ripple Gel Phase of Dihexadecylphosphatidylcholine. Biochem. biophys. res. commun. 174: 102-106.

[145] Ichimori H, Sakano F, Matsuki H, Kaneshina S (2002) Effect of Deuterium Oxide on the Phase Transitions of Phospholipid Bilayer Membranes Under High Pressure. Prog. biotechnol. 19: 147-152.

[146] Komatsu H, Okada S (1997) Effects of Ethanol on Permeability of Phosphatidylcholine/Cholesterol Mixed Liposomal Membranes. Chem. phys. lipids. 85: 67-74.

[147] Komatsu H, Okada S (1996) Ethanol-Enhanced Permeation of Phosphatidylcholine/phosphatidylethanolamine Mixed Liposomal Membranes Due to Ethanol-induced Lateral Phase Separation. Biochim. biophys. acta. 1283: 73-79.

[148] Ahl PL, Chen L, Perkins WR, Minchey SR, Boni LT, Taraschi TF, Janoff AS (1994) Interdigitation-fusion: A New Method for Producing Lipid Vesicles of High Internal Volume. Biochim. biophys. acta biomembr. 1195: 237-244.

[149] Kisak ET, Coldren B, Zasadzinski JA (2002) Nanocompartments Enclosing Vesicles, Colloids and Macromolecules via Interdigitated Lipid Bilayers. Langmuir. 18: 284-288.

[150] Kisak ET, Coldren B, Evans CA, Boyer C, Zasadzinski JA (2004) The Vesosome- A Multicompartment Drug Delivery Vehicle. Curr. med. chem. 11: 199-219.

[151] Zasadzinski JA, Wong B, Forbes N, Braun G, Wu G (2011) Novel Methods of Enhanced Retention in and Rapid, Targeted Release from Liposomes, Curr. opin. colloid interface sci. 16: 203-214.

[152] Raffy S, Teissié J (1997) Electroinsertion of Glycophorin A in Interdigitation-fusion Giant Unilamellar Lipid Vesicles. J. biol. chem. 272: 25524-25530.

[153] Paleos CM, Tsiourvas D, Sideratou Z (2012) Preparation of Multicompartment Lipid-based Systems Based on Vesicle Interactions. Langmuir. 28: 2337-2346.

[154] Chandrawati R, van Koeverden MP, Lomas H, Caruso F (2011) Multicompartment Particle Assemblies for Bioinspired Encapsulated Reactions. J. phys. chem. lett. 2: 2639-2649.

[155] Bolinger P-Y, Stamou D, Vogel H (2008) An Integrated Self-assembled Nanofluidic System for Controlled Biological Chemistries. Angew. chem. int. ed. 47: 5544-5549.

[156] Boyer C, Zasadzinski JA (2007) Multiple Lipid Compartments Slow Content Release in Lipases and Serum. ACS nano. 1: 176-182.

[157] Wong B, Boyer C, Steinbeck C, Peters D, Schmidt J, van Zanten R, Chmelka B, Zasadzinski JA (2011) Design and In Situ Characterization of Lipid Containers with Enhanced Drug Retention. Adv. Mater. 23: 2320-2325.

Permissions

The contributors of this book come from diverse backgrounds, making this book a truly international effort. This book will bring forth new frontiers with its revolutionizing research information and detailed analysis of the nascent developments around the world.

We would like to thank Dr. Amal Ali Elkordy, for lending her expertise to make the book truly unique. She has played a crucial role in the development of this book. Without her invaluable contribution this book wouldn't have been possible. She has made vital efforts to compile up to date information on the varied aspects of this subject to make this book a valuable addition to the collection of many professionals and students.

This book was conceptualized with the vision of imparting up-to-date information and advanced data in this field. To ensure the same, a matchless editorial board was set up. Every individual on the board went through rigorous rounds of assessment to prove their worth. After which they invested a large part of their time researching and compiling the most relevant data for our readers. Conferences and sessions were held from time to time between the editorial board and the contributing authors to present the data in the most comprehensible form. The editorial team has worked tirelessly to provide valuable and valid information to help people across the globe.

Every chapter published in this book has been scrutinized by our experts. Their significance has been extensively debated. The topics covered herein carry significant findings which will fuel the growth of the discipline. They may even be implemented as practical applications or may be referred to as a beginning point for another development. Chapters in this book were first published by InTech; hereby published with permission under the Creative Commons Attribution License or equivalent.

The editorial board has been involved in producing this book since its inception. They have spent rigorous hours researching and exploring the diverse topics which have resulted in the successful publishing of this book. They have passed on their knowledge of decades through this book. To expedite this challenging task, the publisher supported the team at every step. A small team of assistant editors was also appointed to further simplify the editing procedure and attain best results for the readers.

Our editorial team has been hand-picked from every corner of the world. Their multi-ethnicity adds dynamic inputs to the discussions which result in innovative outcomes. These outcomes are then further discussed with the researchers and contributors who give their valuable feedback and opinion regarding the same. The feedback is then collaborated with the researches and they are edited in a comprehensive manner to aid the understanding of the subject.

Apart from the editorial board, the designing team has also invested a significant amount of their time in understanding the subject and creating the most relevant covers. They scrutinized every image to scout for the most suitable representation of the subject and create an appropriate cover for the book.

The publishing team has been involved in this book since its early stages. They were actively engaged in every process, be it collecting the data, connecting with the contributors or procuring relevant information. The team has been an ardent support to the editorial, designing and production team. Their endless efforts to recruit the best for this project, has resulted in the accomplishment of this book. They are a veteran in the field of academics and their pool of knowledge is as vast as their experience in printing. Their expertise and guidance has proved useful at every step. Their uncompromising quality standards have made this book an exceptional effort. Their encouragement from time to time has been an inspiration for everyone.

The publisher and the editorial board hope that this book will prove to be a valuable piece of knowledge for researchers, students, practitioners and scholars across the globe.

List of Contributors

Pratima Parashar
Department of Materials Science, Mangalore University, Mangalagangotri, India CET, IILM Academy of Higher Learning, Greater Noida, India

W. Steinmann, S. Walter, M. Beckers, G. Seide and T. Gries
Institut für Textiltechnik (ITA) der RWTH Aachen University, Aachen, Germany

Luis Alberto Alcazar-Vara and Eduardo Buenrostro-Gonzalez
Instituto Mexicano del Petróleo, Programa Académico de Posgrado. Eje Central Lázaro Cárdenas, México, D.F.

Eliane Lopes Rosado, Vanessa Chaia Kaippert and Roberta Santiago de Brito
Nutrição e Dietética Departament, Federal University of Rio de Janeiro, Rio de Janeiro, Brazil

Kazu-masa Yamada
Hakodate National College of Technology, Department of Electrical and Electronic Engineering, Japan

R. F. B. Gonçalves, J. A. F. F. Rocco and K. Iha
Instituto Tecnológico de Aeronáutica, CTA, São José dos Campos, S.P., Brazil

Daniel Plano
Synthesis Section, Department of Organic and Pharmaceutical Chemistry, University of Navarra, Pamplona, Spain Department of Pharmacology, Penn State Hershey College of Medicine, Hershey, PA, USA

Juan Antonio Palop and Carmen Sanmartín
Synthesis Section, Department of Organic and Pharmaceutical Chemistry, University of Navarra, Pamplona, Spain

Jindřich Leitner, David Sedmidubský and Květoslav Růžička
Institute of Chemical Technology, Prague, Czech Republic

Pavel Svoboda
Charles University in Prague, Faculty of Mathematics and Physics, Prague, Czech Republic

M.D.A. Saldaña and S.I. Martínez-Monteagudo
Department of Agricultural, Food and Nutritional Science, University of Alberta, Edmonton, AB, Canada

Eric A. Smith and Phoebe K. Dea
Department of Chemistry, Occidental College, Los Angeles, USA

Printed in the USA
CPSIA information can be obtained
at www.ICGtesting.com
JSHW011443221024
72173JS00004B/919